ELEMENTARY MODERN CHEMISTRY

AF887928

MAXWELL PRESS

ELEMENTARY MODERN CHEMISTRY

By

Wilhelm Ostwald

Winner of Nobel Prize for Chemistry

Emeritus Professor of Chemistry in the University of Leipzig

And

Harry W.Morse

Instructor in Physics in Harvard University

MAXWELL PRESS

Chennai Trichy NewDelhi

Honour Copyright
&
Exclude Piracy

This book is protected by copyright. Reproduction of any part in any form including photocopying shall not be done except with authorization from the publisher.

MAXWELL PRESS

ISBN 978-93-90063-41-3 Maxwell Press

All rights reserved

Printed and bound in India

No. 44, Nallathambi Street,
Triplicane, Chennai 600 005

MJP 888

© Publishers, 2021

Publisher : C. Janarthanan

ELEMENTARY MODERN CHEMISTRY

By

Wilhelm Ostwald

Winner of Nobel Prize for Chemistry

Emeritus Professor of Chemistry in the University of Leipzig

And

Harry W.Morse

Instructor in Physics in Harvard University

MAXWELL PRESS

Chennai Trichy NewDelhi

Honour Copyright
&
Exclude Piracy

This book is protected by copyright. Reproduction of any part in any form including photocopying shall not be done except with authorization from the publisher.

MAXWELL PRESS

ISBN 978-93-90063-41-3 Maxwell Press

All rights reserved
Printed and bound in India

No. 44, Nallathambi Street,
Triplicane, Chennai 600 005

MJP 888
© Publishers, 2021

Publisher : C. Janarthanan

Publisher's Note

The legacy of a country is in its varied cultural heritage, historical literature, developments in the field of economy and science. The top nations in the world are competing in the field of science, economy and literature. This vast legacy has to be conserved and documented so that it can be bestowed to the future generation. The knowledge of this legacy is slowly getting perished in the present generation due to lack of documentation.

Keeping this in mind, the concern with retrospective acquiring of rare books has been accented recently by the burgeoning reprint industry. Maxwell Press is gratified to retrieve the rare collections with a view to bring back those books that were landmarks in their time.

In this effort, a series of rare books would be republished under the banner, "Maxwell Press". The books in the reprint series have been carefully selected for their contemporary usefulness as well as their historical importance within the intellectual. We reconstruct the book with slight enhancements made for better presentation, without affecting the contents of the original edition.

Most of the works selected for republishing covers a huge range of subjects, from history to anthropology. We believe this reprint edition will be a service to the numerous researchers and practitioners active in this fascinating field. We allow readers to experience the wonder of peering into a scholarly work of the highest order and seminal significance.

MAXWELL PRESS

PREFACE

The beginner in the study of chemistry finds interest and pleasure, first of all, in the remarkable new facts which are placed before him, and this is both natural and desirable. No one can master the science without much study of properties and reactions, and minute acquaintance with as many phenomena as possible is a fundamental requisite. Modern chemistry has, however, a number of general laws which appear everywhere through it, connecting various facts in the most interesting way. Before study has gone very far it becomes necessary to begin the statement of these laws and to indicate how the facts already learned group themselves about them. Two purposes are served by this method of presentation. First, a real *science* of chemistry begins to unfold itself to the student, and second, the acquisition of more new facts is made easier and more interesting.

We have planned in this book to present a sufficient number of facts and experiments to fill the time usually devoted to a first course, and at the same time we have endeavored to fit these facts, as far as possible, to the simpler of the general laws now firmly established as the basis of the science of chemistry.

It is not to be expected or desired that every teacher will wish to follow exactly the order we have chosen. Each teacher of a science must find the individual best way of presenting it. We believe, however, that the experiments chosen and the general laws explained will serve as a fitting introduction, and that they will lead naturally to the desired development of scientific methods of thought by the student.

The drawings illustrating this book were made, in nearly every instance, in the laboratory from apparatus actually set up and tested for performing the experiment described. The successful achievement of this laborious task is due to the scientific training and artistic skill of Miss Elizabeth D. Gray, of Lynn, Mass.

WILHELM OSTWALD
HARRY W. MORSE

CONTENTS

LIST OF FULL-PAGE ILLUSTRATIONS

CHAPTER I

SUBSTANCES AND PROPERTIES

1. Properties. If a body is shown to you with the question, What is this? the answer is to be found by investigating the body in various ways and finding out what its *properties* are. It may be white and made up of little grains with shining surfaces; it may taste sweet and dissolve readily in water, giving a sweet taste to the water; it may make moist fingers sticky when it is handled. From the evidence thus obtained you would conclude that the body is *sugar*.

This conclusion was reached after an examination of the properties of the body. It was observed *by the eye* and found to have a white color and shining surfaces. It was examined by the *taste* and found to be sweet. It was touched and found to be a solid, and by holding it in the hand it could be shown to have weight. When brought into contact with another body, water, it lost its visible properties, disappearing as far as the eye was concerned. But it still retained the properties which belonged to taste, since the liquid formed tasted like the original body.

2. Specific properties. In the study of chemistry we learn all we can about the properties of bodies, but we pay special attention to such properties as are independent of the shape or size of the piece studied. These are called the *specific* properties of the body. *Color* is easy to see — we can apply this test even when we are not very near; *luster* is another such property, and still another is the *state* in which it exists — whether it is a solid, a liquid, or a gas. Our other tests — the taste of the body and its solubility — are harder to apply, and involve the use of other bodies.

3. Substances. A body which is studied with reference to its specific properties is called a *substance.* Sugar is the same substance, whether we mean a large or a small piece.

Experiment 1. Examine a lump of sugar with the eye alone and then with a magnifying glass. Can you find any difference in the specific properties at different points?

Experiment 2. Examine a piece of granite carefully, first with the eye alone and then with a magnifying glass.

The granite evidently consists of more than one substance. The reddish parts with smooth plane surfaces are feldspar, the white parts are quartz, and the black, shiny bits are thin plates of mica.

4. Mixtures. Bodies which have different properties are evidently different substances and we give them different names. If a body consists of several substances which can be distinguished from each other, it is called a *mixture.* Granite is a mixture, since it consists of three substances, — quartz, feldspar, and mica. Sugar is not a mixture but a simple substance, since all the pieces are the same in all of their specific properties. Very often the parts of a mixture are so small that they cannot be distinguished by the eye alone. A magnifying glass or a microscope may then be used in examining them.

Experiment 3. *Demonstration.* An enlarged picture of a section of granite as seen through the microscope should be shown; if possible, as seen by polarized light.

Figure 1 shows the appearance of a section of granite as seen through the microscope. Figure 2 shows the same section illuminated by polarized light.

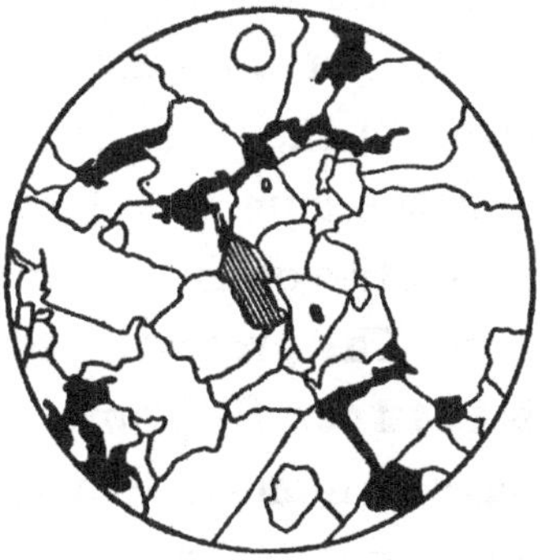

FIG. 1. Section of granite (ordinary light)

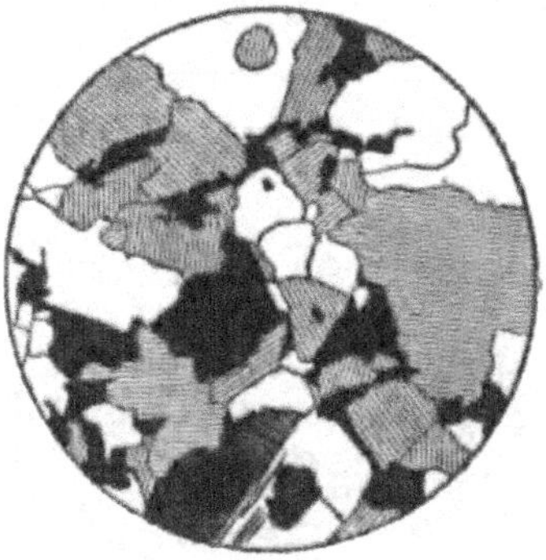

FIG. 2. Section of granite (polarized light)

Many bodies which to the eye are not mixtures are found to consist of several substances when examined with a microscope. The stronger the magnifying power the greater the number of bodies which are found to be mixtures, but even under the most powerful microscope many substances show no parts which are different from one another.

5. Homogeneous substances. Substances which are not mixtures are said to be *homogeneous.* Sugar is a homogeneous substance and so are water and air. The name "substance" is applied to liquids like water and gases like air, as well as to solid bodies like sugar and quartz. All substances have the property of *weight,* which can be felt by holding them in the hand and which can be exactly measured by a balance. As we shall see later, gases have weight also, although they are very much lighter — usually a thousand times lighter — than the same volume of liquids or solids.

In chemistry we study the specific properties of homogeneous substances only. Mixtures, of course, consist of homogeneous substances too, but of two or more different ones; and if a mixture is given to us for study, we separate it into its homogeneous parts and study each separately. We can then add up the properties of the various parts and the sum will be the properties of the mixture.

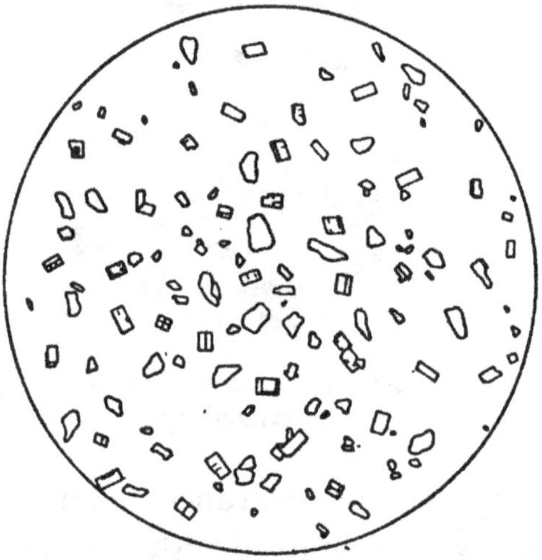

FIG. 3. A mixture of sand and sugar under the microscope

The grains with sharp edges and corners are sugar and the rounded grains are sand

6. The properties of a mixture. Suppose we have a mineral which is composed of equal parts by volume of a black substance and a white one. The color of the mineral will be gray. If the specific gravity of the white substance is 5.0 and that of the black substance is 7.0, the specific gravity of the mineral will be 6.0, for each cubic centimeter of it consists of half a cubic centimeter of a substance of specific gravity 5.0 (that is, 2.5 gm. of it), and half a cubic centimeter of a substance of specific gravity 7.0 (that is, 3.5 gm.), and so the total weight of one cubic centimeter of the mixture will be

$$2.5 \text{ gm.} + 3.5 \text{ gm.} = 6.0 \text{ gm.}$$

7. Phases. The different parts of a mixture (each a homogeneous substance by itself) are called its *phases.* Granite is a mixture of three phases, — quartz, feldspar, and mica. A homogeneous substance consists of only one phase, for every part of it has the same specific properties. Different phases have different specific properties, and when different solid phases are brought together they form a mixture whose properties can be calculated as already shown from the properties of the various phases. With liquids it is not always as simple as this, for they very often act on each other to form a new phase — one which has properties different from those calculated by the simple rule given above. Gases never form mixtures, but unite in every case to form one single new phase. The changes which take place when gases unite are the very simplest of all *chemical processes,* and we shall study about them soon.

QUESTIONS

1. What is a body?
2. What is the distinction between a body and a substance?
3. What is a mixture?
4. What is a homogeneous substance?
5. What is a phase?
6. How can the properties of a mixture be calculated from those of its parts?

CHAPTER II

MELTING AND SOLIDIFICATION

8. Melting point and freezing point. Any solid whatever can be changed into a liquid if its temperature is raised to a high enough point, and any liquid can be changed to a solid if it is cooled to a sufficiently low temperature. Water changes into ice at the freezing point, and ice melts back into water again at the *same temperature.* Water and ice are two phases of the same substance, and when they are mixed together the temperature of the mixture very soon becomes 0° C. This is the freezing point of water and the melting point of ice, and it is the temperature at which ice and water can exist together.

Experiment 4. Place some finely crushed ice in a beaker and watch the temperature on a thermometer placed in the ice. Warm the beaker with the hand. Warm it slightly with a Bunsen burner. At what temperature can water and ice exist in the same vessel?

9. The freezing point of water. A piece of ice always has the same temperature while it is melting, without regard to the amount of water or ice which is present. In the same way, water always has the same temperature while it is freezing. If we keep on cooling it, the water will all change to ice at 0° C. and the solid body, "ice," can then be cooled as far as we choose. If we heat it and keep on adding heat, the ice will all melt at 0° C. and we can then cause the water to rise in temperature; but not indefinitely, since at 100° C. the water will boil away.

All other substances show these same relations; the solid and liquid phases can exist together at only one temperature; this is the melting point if we start with the solid, the freezing point if we start with the liquid. These two points are always at exactly the same temperature, and this is the only point where the solid and liquid phases can exist together.

10. Heat of fusion. We saw from the experiment with the melting ice that it is possible to add a good deal of heat to a mixture of ice and water without changing the temperature of the mixture, provided solid ice is still present. It is just as true that we can take away a good deal of heat from a mixture of ice and water without changing its temperature, provided water still remains. Heat is necessary to melt ice, and heat is given out when water freezes. The quantity of heat which is necessary to melt one gram of ice is called the *heat of fusion of water*; the same quantity of heat is given out when a gram of liquid water freezes.

11. Boiling point. While the solid phase, ice, can be cooled to any temperature without further change, the liquid, water, cannot be heated indefinitely without a change taking place.

Experiment 5. Fill a flask about a third full of water and heat it on a wire gauze or an asbestos plate over a burner. Take readings on a thermometer placed in the water. At what temperature does the water begin to boil? What is the effect of increasing the flame? What becomes of the heat we are adding? Remove the thermometer from the boiling water and let the mercury sink a few degrees. Then bring it back into the vapor just above the water, but not touching it. What is the temperature at this point?

Water is, of course, a substance, and it is one of a very large class which behave toward changes of temperature in the same way. If we heat the solid, the temperature rises to a certain point and then remains stationary until all of the solid phase is changed into the liquid phase. If we heat the liquid phase of one of these substances, the temperature rises to a certain point and then remains stationary until all of the liquid has been changed into vapor.

12. The three "states." While ice, water, and steam are three different substances, with very great differences in some of their specific properties, they are in chemistry usually considered as the same material, because of the ease with which one can be completely changed into the other, and also because when they react with other substances they give the same result. Ice is therefore often called "water in the solid state," and steam "water in the gaseous state."

Every substance known is either solid, liquid, or gaseous, and there is a set of general properties which enable us to decide where a substance belongs. Solids retain their shape, which is a very important property indeed, and one without which the present conditions of things on this earth could not continue. The word "solid" is a description of the fact that the shape is retained. Liquids have no definite shape of their own under ordinary circumstances. They take the shape of any vessel in which they are placed. (How about small drops?) But liquids and solids both have the definite property of volume. Gases have no shape of their own, nor have they any definite volume, but expand to fill the entire volume of any vessel into which they are brought. Whether a substance exists in one of these states rather than in another depends on the temperature and, as we shall see later, in some degree on the pressure also.

13. Range of existence of the three "states." There are, of course, liquids with very low freezing points, and solids with very high melting points. A solid substance can exist at as low a temperature as we know how to produce, but it is limited in upward range by the melting point of the solid. A liquid exists between its freezing point and its boiling point only. A gas has as lower limit its point of condensation, but has no upward limit of existence in temperature.

14. Sublimation. A few substances pass from the form of a solid directly into that of a gas, without melting; they are said to *sublime.*

Experiment 6. Heat some ammonium chloride in a hard glass test tube.

QUESTIONS

1. At what temperature can water and ice exist together?

2. Explain how the freezing of a pond keeps the temperature about it higher than it would be if no water were there.

3. What are the general differences in the properties of solids, liquids, and gases?

4. What is sublimation?

CHAPTER III

PURE SUBSTANCES AND SOLUTIONS

15. Properties. All substances do not behave as pure water does in the matter of melting and freezing. We often speak of sea water simply as "water," but if its specific properties are compared with those of pure water, they are found to be different. It is easier to swim in the ocean than in a lake because sea water is denser than fresh water. Its boiling point is higher and its freezing point is lower than the corresponding points for fresh water.

The properties of pure water are always the same, no matter how it is obtained, but the properties of different samples of sea water are found to vary. Near the mouth of a river sea water is nearly like fresh water, and by mixing sea water with varying proportions of fresh water an unlimited series of very similar liquids can be made, with properties varying continuously from those of pure water to those of the strongest sea water.

Pure water boils away without leaving any residue, but sea water leaves, on evaporation, a crust of sea salt. If this salt is put into pure water it disappears, or *dissolves*, and a liquid is formed which has the same properties as the original sea water.

There is another very striking difference between sea water and fresh water. The boiling point of the former is not only higher than that of the latter, but it is *not constant*, and the longer the sea water has been evaporating the higher the temperature will appear on a thermometer in the boiling liquid. The freezing point changes in just the same way; as more ice is formed in the liquid the temperature gets lower and lower.

Experiment 7. Fill a beaker about one third with a 5% solution of common salt in water. Boil the solution and observe the boiling point on a thermometer at intervals of five minutes.

Experiment 8. In a mixture of ice and salt cool a beaker one third filled with a 5% salt solution. What is the freezing point of the solution? Continue to cool after ice has begun to separate, taking occasional readings of the thermometer.

16. Solutions. There are two distinct classes of homogeneous substances: those with constant properties are called *pure substances* and the others are called *solutions.* Every homogeneous substance is either a pure one or a solution, and a test of its properties will always decide to which class it belongs.

17. The boiling point of solutions. The boiling point of a water solution may be higher or lower than that of pure water, depending on the nature of the other substance which is present in the solution. If this has a lower boiling point than water, then the solution will generally have a lower boiling point. Solutions made of a substance with a high boiling point will have a higher boiling point than water.

Experiment 9. Try the boiling point of a solution of alcohol in water.

18. The freezing point of solutions. While the boiling point of a water solution may be either higher or lower than that of pure water, the freezing point of a solution is always lower than that of water.

Experiment 10. Try the freezing point of a solution of alcohol in water.

Experiment 11. *Demonstration.* Show the freezing of a strong solution of alcohol with a mixture of solid carbon dioxide and ether. Liquid carbon dioxide may be had of any dealer in chemical supplies.

19. The name "solution." The name "solution" is given to homogeneous bodies with changing boiling and freezing points, because they can be made by "dissolving" one body in another; that is, by bringing together two phases which act on each other to produce a single phase. There are three kinds of solutions, — solid, liquid, and gaseous. Solid solutions we shall meet with but rarely, so we will not study about them here. Gaseous solutions are common and we shall learn about them in a later chapter. *Liquid solutions* are very common indeed, and they are made by dissolving either a solid, a liquid, or a gas in a liquid.

Experiment 12. Try the solubility of a number of solids in water, alcohol, and kerosene. Sulphur, salts from the laboratory shelves, rubber, etc., may be tried. Try the solubility of a number of liquids in each other, — water, alcohol, carbon bisulphide, ether, and kerosene, for example. *All flames should be extinguished* during these experiments.

Experiment 13. Try the solubility of carbon dioxide gas in water. Have you any evidence that air dissolves in water?

20. The properties of solutions. The properties of solutions made from the same substances may vary widely; in fact, all the way from one of the substances used to make the solution, to the other.

Experiment 14. Make a concentrated solution of potassium permanganate. With how many times its volume of pure water can you dilute it before the color disappears?

21. The separation of solutions into components. Every solution can be separated into two or more pure substances, and one of the most usual methods of separation is by boiling.

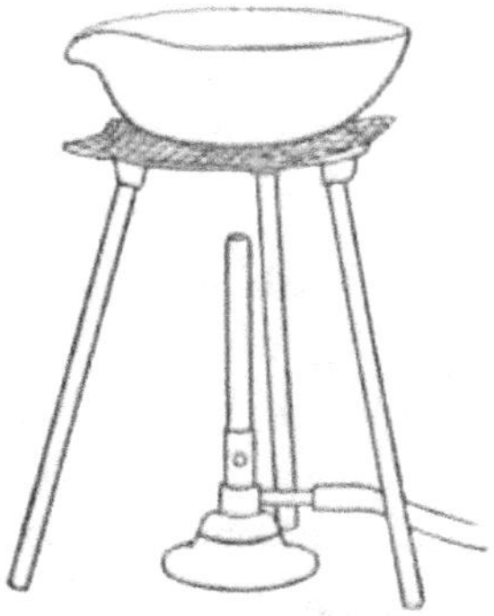

FIG. 4. Apparatus for evaporating the water from a solution

Experiment 15. Evaporate to dryness a solution of common salt in water in an evaporating dish over a burner. What becomes of the water?

Cool a bottle of soda water on the ice and open it while it is cold. Pour it into a beaker and warm it with the flame.

We have separated these solutions into their components. When we evaporated the salt solution, what escaped in the steam was *pure* water and what was left behind was *pure* salt. As we shall see in the chapter on water, it is not difficult to catch the water which goes off and get it back into the form of liquid pure water.

FIG. 5. Freezing out ice from a solution

Another way of making this separation is by freezing.

In the first case we made the separation by changing the water into vapor, another form of water, and in the second we made the separation by changing it into ice. In most cases the first method is the more convenient.

Experiment 16. *Demonstration.* Partly fill a beaker with a solution of potassium permanganate (0.2 gm. in 200 cm. of water). Make a freezing mixture of ice and salt, and fill a small metal vessel, a calorimeter, for example, with it. Place the vessel containing the freezing mixture in the permanganate solution and let it remain until about an eighth of an inch of ice is frozen out. Show the difference in color between the solution produced by melting this ice and the original solution.

QUESTIONS

1. How does the boiling point of a solution of salt in water differ from the boiling point of water?

2. How does the freezing point of such a solution differ from the freezing point of water?

3. What about the boiling point of a solution of alcohol and water?

4. What about the freezing point of such a solution?

5. Make a table of solubilities from the data of your experiments.

6. When the ocean freezes is the ice fresh or salt?

7. What are the general methods of separating a solution into its component parts?

CHAPTER IV

COMBUSTION

22. Phenomena. When a candle burns, light and heat are produced and the candle gradually disappears. At first glance the candle seems to be the only thing that is necessary, but a simple experiment will show that a supply of air is needed also.

FIG. 6. Burning out part of the air from a beaker

Experiment 17. Stick a candle to a small, round piece of board and float it on water in a basin. Invert a large beaker over board and candle, as in Figure 6.

Some gas is left in the beaker after the candle stops burning. It is evident that this gas is not air, for one of the specific properties of air is to help a candle to burn.

Experiment 18. Fasten a candle to a wire and lower it into a large flask. After it has gone out, remove it from the flask, light it, and put it back into the flask.

Experiment 19. Pour a little limewater into the flask where the candle has been burning. Pour some limewater into a flask in which no candle has been burning.

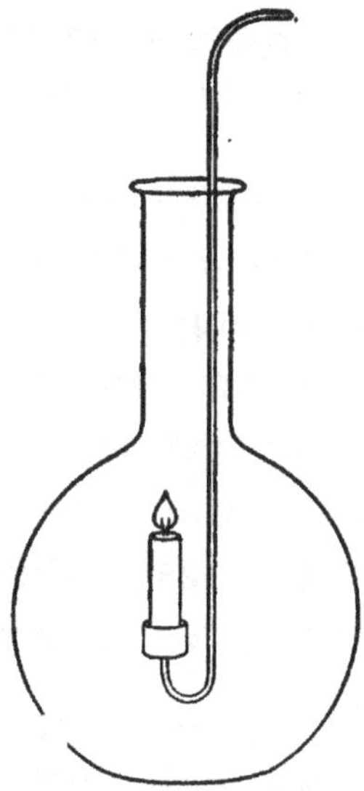

FIG. 7. Burning out the combustible part of the air in a flask

23. Reagents. The gas in the flask in which the candle burned has properties which are not like those of air. Like most gases, its properties are not easily recognized by our senses. It looks like air and has no color, odor, or taste, but the limewater shows that it is not air. The limewater is a ***reagent*** which is used to detect something not directly appreciable by our senses. We shall learn in chemistry to use a great many reagents as aids in distinguishing various substances, and every effect produced by the use of such a reagent gives us a new additional specific property of the substance examined.

FIG. 8. Collecting water which has been formed by combustion

24. The products of combustion. What becomes of the candle when it burns?

Experiment 20. Hold a cold, dry beaker inverted over the burning candle. What forms on the glass?

Water is one of the products formed when a candle burns, but it is evident from the change produced in the limewater that something else beside water was there. This was a gas, which is called *carbon dioxide.*

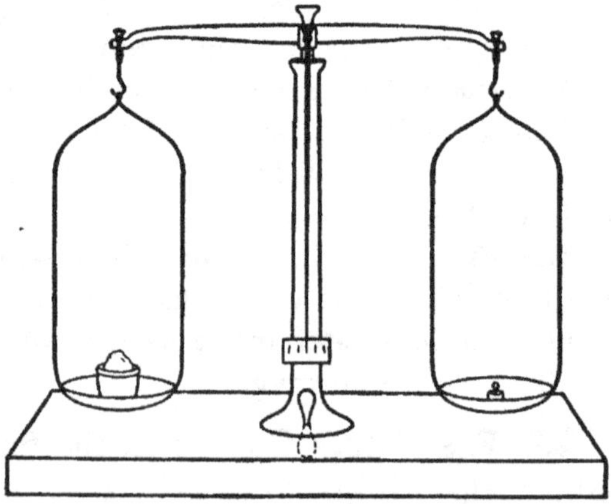

FIG. 9. The proof that iron gains in weight when it burns

Experiment 21. Support a crucible lid on one pan of a balance on a small iron tripod. Pour a little heap of reduced iron into the lid and balance by adding weights to the other pan. Light the iron powder with a flame and state results.

In this case the product of combustion is solid.

Experiment 22. Test the magnetic properties of this solid. Is it iron? Is its weight greater or less than that of the iron used in the experiment?

Experiment 23. Heat weighed amounts of zinc dust and fine copper filings to a high temperature in uncovered porcelain crucibles. Then cool and weigh again. Is anything absorbed or given off?

When a gas is produced during combustion it is not so easy to catch and weigh it. There are fortunately substances which absorb and hold water and carbon dioxide. Solid caustic soda is such a substance.

Experiment 24. Arrange apparatus like that in the figure. The upper part of the cylinder contains caustic soda supported on a diaphragm of wire gauze. Place this on one pan of a balance, balance by weights on the other pan, and light the candle. Then state your results.

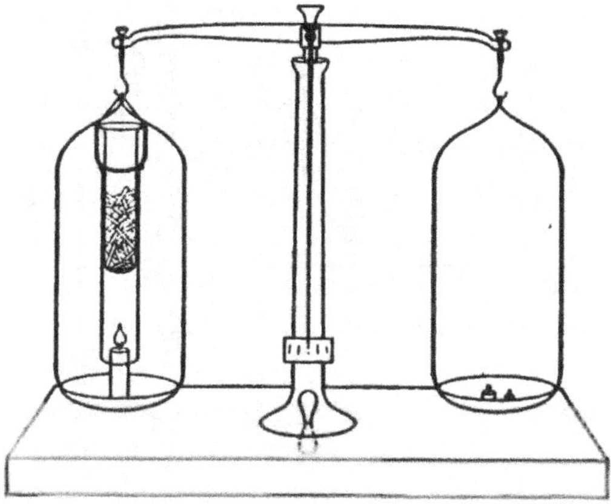

FIG. 10. The detection of one of the invisible products of combustion by weighing it

From all of our experiments on combustion we may conclude that the bodies produced always weigh more than the original bodies did. It seems also to be evident that air is necessary, but the experiment with the candle (Exp. 17) indicates that the air is not all used up during the process.

Experiment 25. Repeat Experiment 17, but in place of the candle use a small heap of sulphur in a crucible lid supported on an iron tripod. Ignite the sulphur with a match and cover with the beaker. What fraction of the air is used up?

25. Oxygen. Less than a quarter of the air takes part in the combustion. Air is a solution of two gases, and the one which is active in helping things to burn is called ***oxygen.*** The other part is called ***nitrogen*** and we shall study more about it later (Chap. VIII).

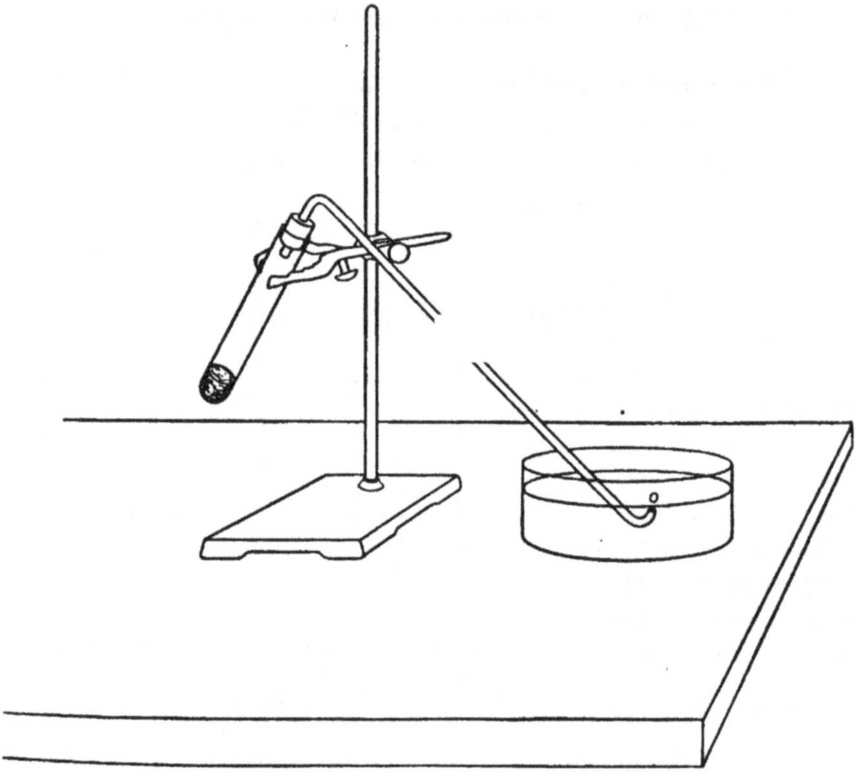

Fig. 11. Apparatus for making oxygen from the red oxide of mercury

If ordinary air is used, only the oxygen takes part in a combustion and the nitrogen is left behind. If we had pure oxygen instead of air, there would be no gas left, provided we used enough of the other substance to take up all the oxygen. Some products of combustion give up their oxygen very readily.

JOSEPH PRIESTLEY (1733–1804)

Discovered oxygen, carbon monoxide, and nitric oxide. Investigated the properties of sulphur dioxide, hydrochloric acid, and ammonia as gases. Invented the pneumatic trough

Experiment 26. Weigh a hard glass tube sealed at one end. Add two or three grams of the red oxide of mercury and weigh again. Heat over a burner, holding almost horizontally during the heating. Weigh the tube again after it has cooled and examine the substance remaining in it. What remains? Did anything visible escape?

Experiment 27. Fit a tube like the one used in the previous experiment with a cork and delivery tube. Use about five grams of the red oxide. Heat, and by means of a *pneumatic trough* collect the gas given off (see Fig. 11).

The gas produced is now to be tested for its specific properties. None of these are very evident, for it is colorless, odorless, and tasteless. We must find a reagent which will tell us more about it.

Experiment 28. 1. Test one of the bottles of gas with a glowing splinter of wood.

2. Fasten a piece of charcoal to a wire; ignite it and put it into one of the jars of gas.

3. Try the same experiment with a piece of sulphur in an iron spoon.

4. Try the same with a small piece of phosphorus. (CAUTION!)

5. Try it with a thin coil of iron wire to the end of which the head of a match has been tied.

NOTE. Oxygen for these experiments should be furnished by the laboratory. The teacher may make it from potassium chlorate (see p. 117), or it may be bought in compressed form in cylinders (see Fig. 12).

FIG. 12. Cylinder for compressed gases

26. Combustion in oxygen. Substances burn much more rapidly in oxygen than in air. The same quantity of heat is produced when a gram of a substance is burned, whether in air or oxygen, but in the latter case only the products of combustion have to be heated, while in air all the nitrogen has to be heated too.

The temperature of combustion in oxygen is therefore very high, and the bright light is an indication of this fact. There is a very general law which may be stated here: *A chemical reaction takes place more rapidly the higher the temperature.* The heat and light which are to be seen during these changes are not substances, for all substances have weight, and light and heat have none.

27. Energy. Heat and light are forms of *energy*, and we say that a set of substances have chemical energy when they can react with each other to form new substances, giving out heat or light or some other form of energy during the process.

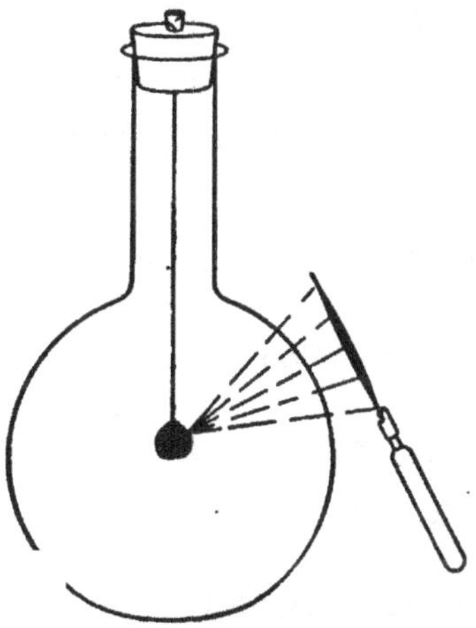

FIG. 13. Apparatus used to show that there is no total gain in weight during combustion

28. Constituents and compounds. In Experiment 26 oxygen gas was made by heating the red oxide of mercury; the gas escaped and the mercury remained behind in the cooler part of the tube. From one substance, mercuric oxide, two others, mercury and oxygen, were made, and it is possible to cause these two to unite again to form mercuric oxide. Mercury and oxygen are the *constituents* of mercuric oxide, and mercuric oxide is a *compound* of mercury and oxygen.

Experiment 29. Fasten a piece of charcoal to the end of a wire and place it in the center of a flask of oxygen which is tightly stoppered. Place the whole on a balance and add weights to the other side until equilibrium is produced. Ignite the charcoal by means of the sun's rays concentrated by a burning glass. Is there any gain in weight during the combustion?

NOTE. In case bright sunlight is not available, the charcoal can be ignited by means of a thin platinum wire heated to redness by a cell or two of storage battery or by a few dry cells.

29. The conservation of weight. This experiment gives us a clew to one of the most important of all physical and chemical laws: *Whatever changes take place between a set of substances, their total weight is never changed.* The individual weights do change, but the sum always remains constant.

By making use of this law we can calculate the relations by weight in which substances enter into chemical processes without weighing every substance by itself. If we use a weighed amount of mercuric oxide and weigh the mercury produced by its decomposition, then we can calculate exactly the weight of the oxygen which escaped. The formula

mercuric oxide = mercury + oxygen

expresses what takes place and includes all the substances that took part in the reaction; and if we can write instead the

weight of mercuric oxide = the weight of mercury
+ the weight of oxygen,

we can calculate the weight of the oxygen without collecting and weighing it.

30. The weight of gases. Gases have weight, as can be shown by the following experiment.

Experiment 30. Weigh a strong glass flask provided with a stopcock. Connect the flask with a valve and a bicycle pump. Open the stopcock and after pumping for a few strokes close it again. Weigh the flask again. Then by means of a delivery tube (see Fig. 14) allow the air which has been pumped in to escape into a flask over the pneumatic trough. Mark the level of the gas in the collecting bottle and then determine the volume of the air by inverting the bottle, filling it with water up to the mark, and measuring the volume of the water in a measuring cylinder. What is the density of air?

31. Law of constant proportions. One of the characteristic properties of a substance is that a definite amount of it can always be transformed into a definite amount of another substance. This means that the quantity of oxygen which can be obtained from two grams of mercuric oxide is just twice the quantity which can be obtained from one gram. This statement is called the *law of constant proportions.*

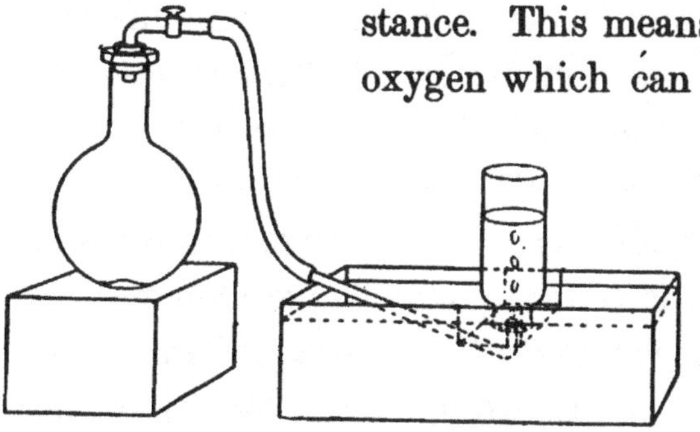

FIG. 14. Apparatus for the determination of the density of air

32. Elements. From what we have already learned it is evident that either mercury or oxygen weighs less than the mercuric oxide from which they were produced. A constituent of a compound always weighs less than the compound itself. A great many experiments have been made with oxygen, but no substance has been produced from it which weighs less than the weight of oxygen used in making it. We can say that *oxygen forms only compounds* and that no constituents of oxygen are known. Such a substance is called a *chemical element.* Mercury, silver, iron, and phosphorus are elements, and there are about seventy-five which we know at present. In the front cover of this book is a table of the elements with the symbols which are used in writing chemical formulæ, and with a number belonging to each element, which we shall learn about soon. Only a few of the elements in this table are at all common, and some of them are very rare indeed. That means that their compounds occur only in very small quantities in the earth's crust.

33. The conservation of the elements. Not only is it impossible to break an element up into constituents, but it has so far been found impossible to change one element into another. This is the same as saying that no amount of an element can ever be created or destroyed. This statement is called the *law of the conservation of the elements.* A few hundred years ago chemists spent their whole lives in trying to change common elements, lead, for example, into rarer and more valuable ones, usually into gold. These alchemists, as they were called, believed that it was possible to produce such a change, and it was only after many years of fruitless effort in this direction that this law of the conservation of the elements was finally stated and believed. Like all other natural laws it is purely a statement of experience. Every one who has tried it has found it to be true.

This law means that it is not only impossible to change one element directly into another, but also that it is impossible to do it indirectly. If we combine three elements, A, B, and C, into compounds in all sorts of ways, make them react together as we like, and then break them up again into constituents, we never get out anything but the three elements, A, B, and C, in exactly the same amount as we started with, provided, of course, that we did not lose anything or add any other substance during the experiments. Or if we find that one compound can be resolved into the elements A and B and another into the elements C and D, no possible reaction between these compounds will ever give anything which cannot be resolved into the elements A, B, C, and D again. The elements act as though they contain something unchangeable, which persists in all of their compounds and can be extracted from them without loss.

34. Elementary composition. Every element, therefore, forms a certain limited group of compounds which are said to *contain* it, because it can be obtained from them unchanged. These groups are like the families of the various elements, and of course every compound belongs to as many families as it contains different elements. It is possible, then, to find for every compound an *elementary composition*, that is, what elements it contains and how much of each. This elementary composition is quite independent of the way in which the compound is prepared. Sugar, for example, is made from sugar cane and from beets, but if it is *pure sugar* its elementary composition is exactly the same, whatever its origin may have been.

QUESTIONS

1. In order that a candle may burn, what is necessary beside the candle itself?

2. What becomes of the candle when it burns?

3. What is a reagent?

4. How would you define the term "combustion"?

5. In the experiments of this chapter do the products of combustion weigh more or less than the substance burned?

6. If you include air as one of the substances which takes part in each of these combustions, what would be your answer to this question?

7. State what you have learned about the properties of oxygen.

8. What other general facts do you know already about the properties of oxygen?

9. If a bright light is produced during a chemical reaction, what is your natural conclusion about the temperature in the reacting mass?

10. How does change of temperature affect the rapidity of chemical action?

11. What are the constituents of mercuric oxide?

12. What is a compound?

13. State the law of the conservation of weight.

14. What is the density of air? How was it determined?

15. What is the weight of one liter of air?

16. State the law of constant proportions.

17. What is a chemical element?

18. State the law of the conservation of the elements.

CHAPTER V

THE METALS

35. Light and heavy metals. For the sake of convenience the chemical elements are divided into two classes, — metals and nonmetals, — and the metals are again divided

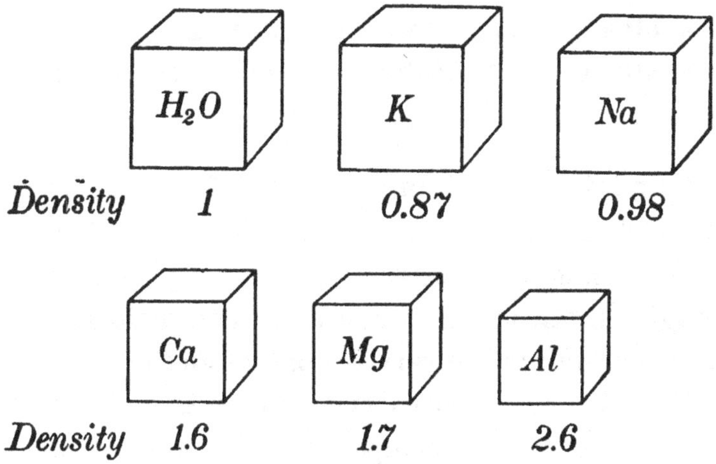

FIG. 15. Specific volumes and densities of the light metals

The cube shows the volume occupied by one gram of each metal; the numbers below are the corresponding densities

into light and heavy metals, the light ones being those having a density less than four times that of water. The metals have a set of properties which are common to all of them in a greater or less degree. They are usually heavier than the nonmetals. They have a surface luster which is called "metallic luster." They are good conductors of heat and electricity, and are usually ductile and malleable. They have in addition some very definite chemical properties which we shall learn about later.

Among the light metals there are five which are of special importance because they are so useful in human affairs. These are

the alkali metals, — sodium, potassium;
the alkaline-earth metals, — magnesium, calcium;
the earth metals, — aluminium.

36. Aluminium. Aluminium has all of the properties common to metals except weight. Its density is very low — only about 2.6 times that of water. It is silvery white in color, has a bright luster, and is very ductile and malleable. Compounds of aluminium are very common indeed on the earth's surface, for almost all rocks and soil contain this metal in the form of its oxygen compounds. In tropical lands the clays are very often nearly pure aluminium oxide.

The element itself does not occur on the earth because it combines very readily with oxygen, and in fact it takes a great deal of work to separate the metal from its compounds. The difference in price between a pound of aluminium, which costs about forty cents, and the same weight of aluminium in the form of clay, which costs almost nothing, is a measure of the amount of energy contained in the element aluminium compared with the amount contained in the oxide. During the past few years we have learned to make metallic aluminium by means of an electric current, and this gives an economical method of furnishing the large amount of energy necessary for the process. When aluminium burns with oxygen to form the oxide, the same amount of energy is given out as was used up in forming the metal from its oxide, but in this case the energy is not given out as electrical energy, but in the form of heat and light. Most photographic flash powders are nowadays made of finely powdered aluminium mixed with a substance which can furnish the oxygen necessary for the combustion.

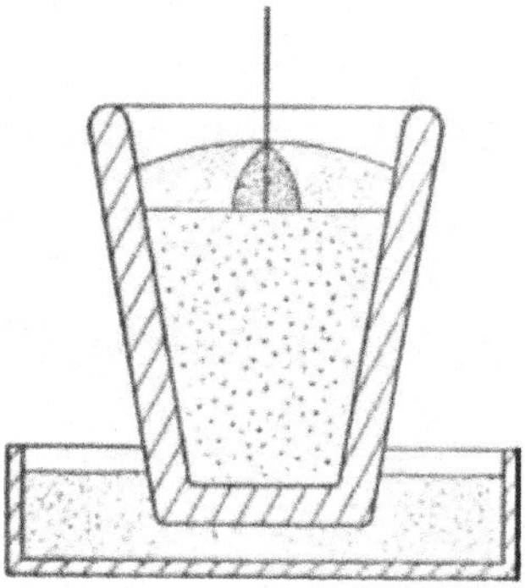

FIG. 16. Apparatus for making iron from thermite

The crucible is supported in a tray of sand to avoid burning the table

Experiment 31. *Demonstration.* Make metallic iron from its oxide by the use of "thermite." (This is a mixture of metallic aluminium and iron oxide, which may be had from many of the chemical-supply houses. A list of interesting experiments may be had from the manufacturer.)

Aluminium has the power of combining with the oxygen of many metallic oxides, burning to aluminium oxide and leaving the other metal, which was previously combined with oxygen, in the free state.

37. Magnesium. Magnesium is a white metal, even lighter than aluminium. It burns in the air with a very bright flame. Magnesium wire or ribbon can be kindled with a match, the temperature produced in this way being high enough to cause combustion to begin. As it goes on, the temperature of the wire near the part which is burning remains high enough to keep the process going. The white smoke which forms is a compound of magnesium and oxygen — called magnesium oxide. Magnesium compounds occur in great quantities on the earth. The metal is a constituent of the mineral dolomite, of which some quite large mountain ranges consist, and compounds of magnesium are common in nearly all rocks.

The burning of magnesium in air, with the production of a high temperature, as indicated by the brilliant light produced, suggests that magnesium is like aluminium in requiring a great deal of work for its production from its compounds. This is the case, and it is in just the same way the reason for the high cost of the metal as compared with that of its compounds.

38. Calcium. This is another light metal whose compounds are very common indeed on the earth, but which

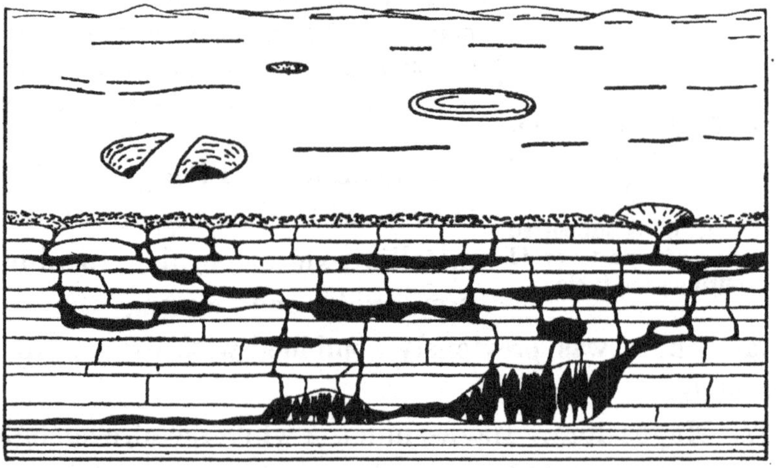

FIG. 17. Section through a region of limestone caves

requires the expenditure of even more energy than is necessary in the case of aluminium and magnesium to separate it from its compounds. It therefore burns still more easily and with the production of more heat than these other metals. Within the past two or three years chemists have learned how to make calcium by means of the electric current in ways which give far better results than those formerly in use. As a result the price of metallic calcium has dropped within two years from about thirty dollars an ounce to about ten cents.

Limestone is a compound containing calcium, and it sometimes makes up whole ranges of mountains. Chalk and marble are the same compound in a different form. Of course all the properties of limestone, chalk, and marble are not the same. They differ in fineness of grain and in other ways, but if they are examined by chemical reagents they are found to give the same reactions, and this is the same as saying that their chemical properties are the same.

39. Sodium. Sodium is a very light metal which combines with oxygen even at ordinary temperatures. When a piece of sodium is cut, the surface retains its metallic luster for only a very short time, becoming immediately coated with a gray layer of oxide. Metallic sodium is always kept away from the air, either in some liquid which does not contain oxygen or else in a tightly soldered metal box.

Sodium combines with oxygen so readily that it takes the oxygen out of many of its compounds even at ordinary temperatures, and when we come to the study of the element hydrogen we shall make use of this property in a very remarkable way.

The most ordinary compound of sodium is common salt — chloride of sodium. Sodium occurs on the earth in Chili saltpeter, borax, and in many other compounds. *Potassium* is very much like sodium, and its compounds occur in great quantities in nature. The feldspars, which are constituents of almost all granites, contain potassium in the form of some of its compounds, which are washed out of these rocks by rain and carried to the plants of the earth, nearly all of which need potassium to live upon.

40. The heavy metals. Copper, gold, tin, lead, silver, and iron are the ones which have been longest known, because they occur in the form of metals in the earth, or can be very easily produced from their compounds. Iron was the last of

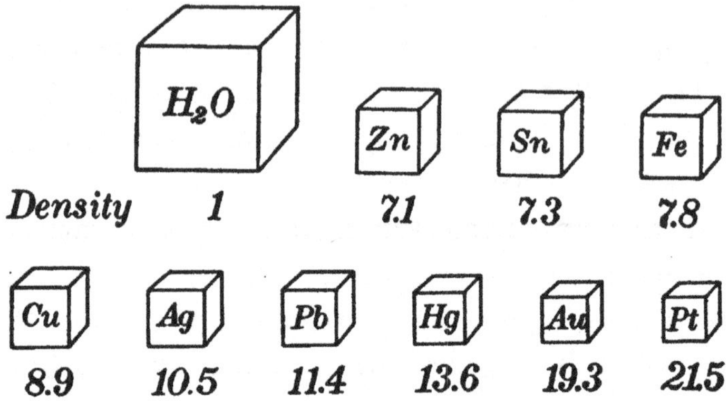

FIG. 18. Specific volumes and densities of the common heavy metals

these to be made, as its production in the metallic state is much more difficult than in the case of the other metals mentioned.

QUESTIONS

1. What are the general properties of a metal?
2. Which are the five most important light metals?
3. Why are they important?
4. What are the properties of aluminium?
5. Why does metallic aluminium not occur in nature?
6. Why is the cost of metallic aluminium so high?
7. State what you know about calcium and magnesium.
8. How does sodium commonly occur in nature?
9. What heavy metals have been known for a very long time?
10. What determined the order in which they became known?

CHAPTER VI

WATER

41. Natural waters. About five sevenths of the earth's surface is covered with water, as we learned when we studied geography, but the water of our seas, lakes, and rivers is not

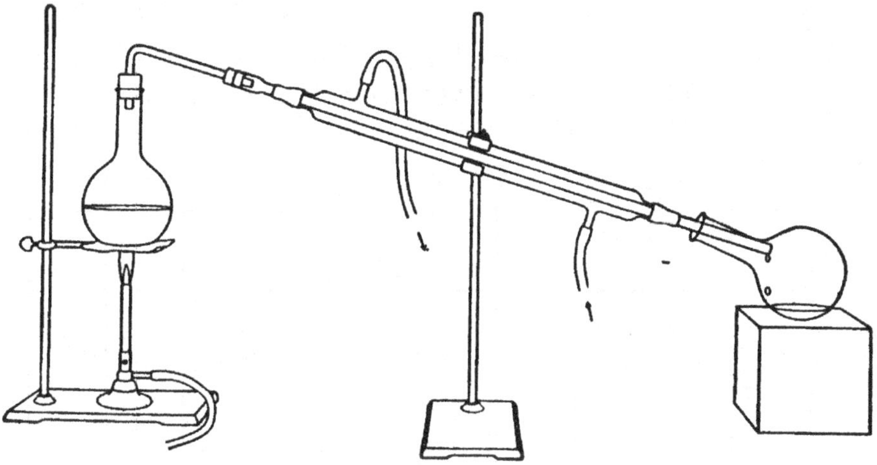

FIG. 19. Apparatus for distillation and condensation

a pure substance. It is a solution, and you will find that pure water has a different taste from that of natural waters, and that it leaves no residue on evaporation, as ordinary water always does.

Experiment 32. Arrange an apparatus like the figure and fill the boiling flask about a third full of water to which a little ink has been added. Boil the water and examine the distillate, as the part is called which collects in the cooled receiver.

42. Distilled water. In order to carry on this process of distillation we need first of all to supply heat to the boiling flask — enough to raise the temperature of the water to boiling and to continue the evaporation of the water as long as we choose. We must also arrange to take away heat from the water vapor, so that it will condense to liquid water again. The burner gives a constant supply of heat and the condenser is used to take the heat away from the water vapor. The

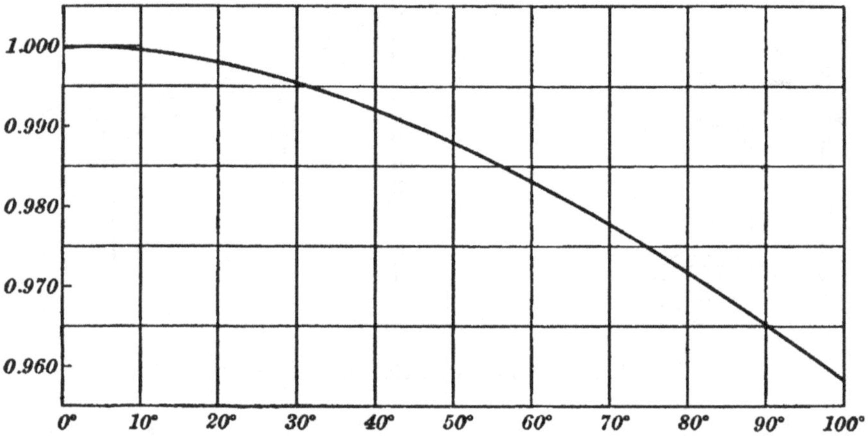

FIG. 20. Curve of density of water from 0° to 100° C.

cooling water is allowed to enter at the bottom of the condenser because the cooling power of the water is used to the best advantage by this arrangement. The steam which enters at the top of the condenser meets warmer water first and then colder and colder water as it passes down the tube. There is no mixing of hot and cold water in the cooling tube of the condenser, since the warm water rises to the top because of its smaller density.

Experiment 33. Examine the properties of distilled water.

43. The maximum density of water. Most substances expand on being heated and contract on being cooled. Water is a very remarkable exception to this general rule, for it has its greatest density at 4° C. and a less density at all other temperatures. This means that water contracts when it is heated from 0° C. up to 4° C. and expands when it is cooled from 4° C. down toward the freezing point. It expands again on freezing, and ice has a much lower density than liquid water. These facts are of the greatest importance in nature.

When still water, like that in a lake, is cooled at its surface, the denser cold water sinks and the lighter warm water rises until the whole lake has reached the temperature of

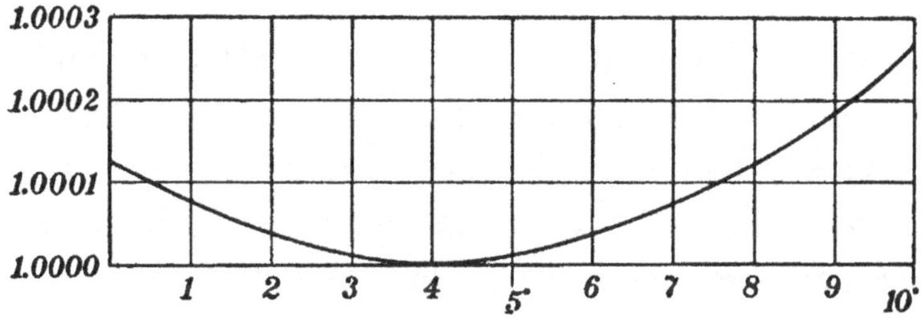

FIG. 21. Curve of the specific volume of water at temperatures in the neighborhood of 4° C.

4°C. As the lake is cooled down still more the colder water is now lighter and remains at the surface until it freezes, while the temperature deep down in the lake still remains about 4° C. This is why bodies of water always freeze over on top first, the ice gradually getting thicker as the cold weather continues. If water were like almost all other liquids, increasing in density with decrease of temperature and having a greater density in the solid form than in the liquid form, ponds and lakes and seas would freeze at the bottom first, and long-continued cold weather would freeze up all fish and other life in a solid mass of ice. The fact that water has its maximum density at a temperature above the freezing point determines many of the cold-weather effects of our winter. Spring would be very different if ice were not lighter than water, and we should have to wait for it until the ponds thawed down to the bottom.

44. Supercooling. Any liquid can be cooled below the freezing point without the formation of the solid phase, provided care is taken in the experiment.

Experiment 34. Mix some pounded ice or snow with salt and stir the mixture with a test tube partly full of water in which is placed a thermometer. What temperature does the water reach before ice appears?

The water is said to be *supercooled,* and this experiment gives us an idea for the following general law: *Even though the conditions necessary for the appearance of a new substance may be all fulfilled, the substance itself does not necessarily appear immediately.* We shall see further proof of this law every day in our chemical work.

Experiment 35. Drop a very small crystal of ice into the supercooled water. What happens? Watch the thermometer as well as the water. Has the water all turned to ice?

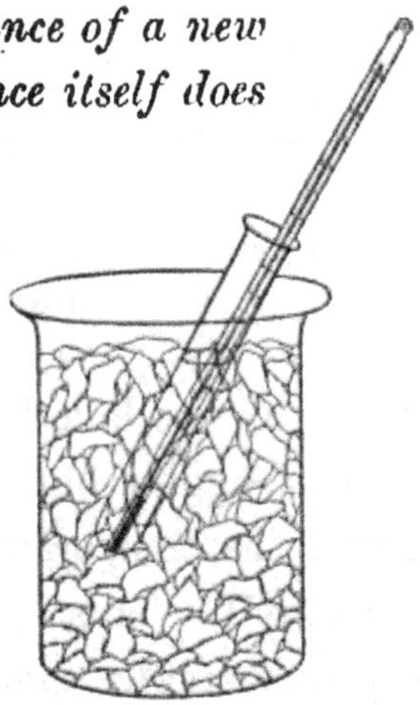

FIG. 22. Cooling water below its freezing point

A very common sight in winter is a small pond which is partly frozen over, and if the supercooled water in the previous experiment is examined immediately after the formation of ice, it will be found that not all of the water is frozen. As we have found (Exp. 4), it takes a good deal of heat to melt ice, even without changing its temperature, and we get the same amount of heat back when water freezes to ice.

The freezing of the water in the tube produced so much heat that the temperature of the mixture rose to the freezing point of water. If any more ice had formed, the temperature would have gone above the freezing point, so of course no more ice could form.

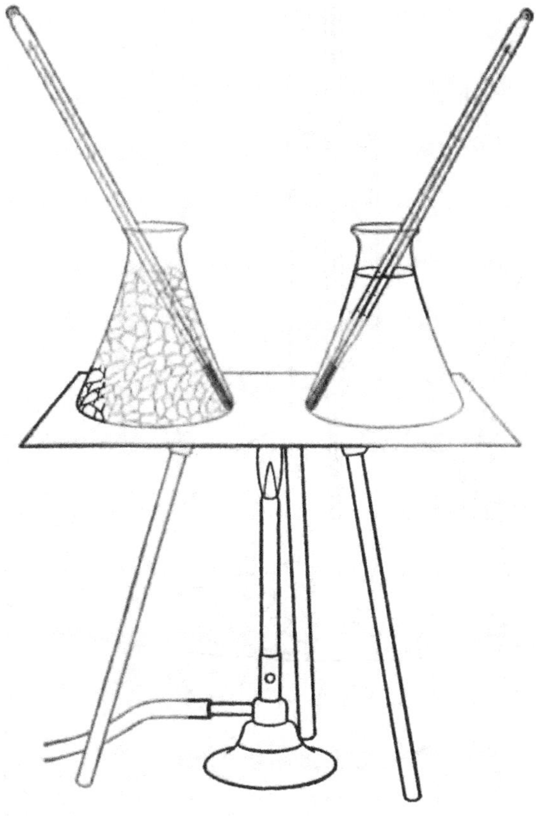

FIG. 23. Apparatus for showing the latent heat of melting of water

45. The heat of fusion of water. The amount of heat required to change one gram of ice at 0° C. to water at 0° C. is called the *heat of fusion* of water. The same amount of heat is given out when one gram of water at 0° C. changes to ice.

Experiment 36. On a flat iron plate over a Bunsen burner place two beakers, one containing a weighed amount of ice and the other an equal amount of water which has been cooled to 0° C. Arrange these beakers on the plate symmetrically, so that each gets the same amount of heat from below. Place a thermometer in each beaker and watch the two thermometers while the plate is heated.

46. The boiling point. Under ordinary conditions water boils at 100° C. It is easy to show that it may boil at other temperatures lower than this.

Experiment 37. Boil some water in a good-sized flask. When it is boiling strongly, remove the flame and then stopper the flask tightly. After it has cooled for a minute or two remove it from the tripod

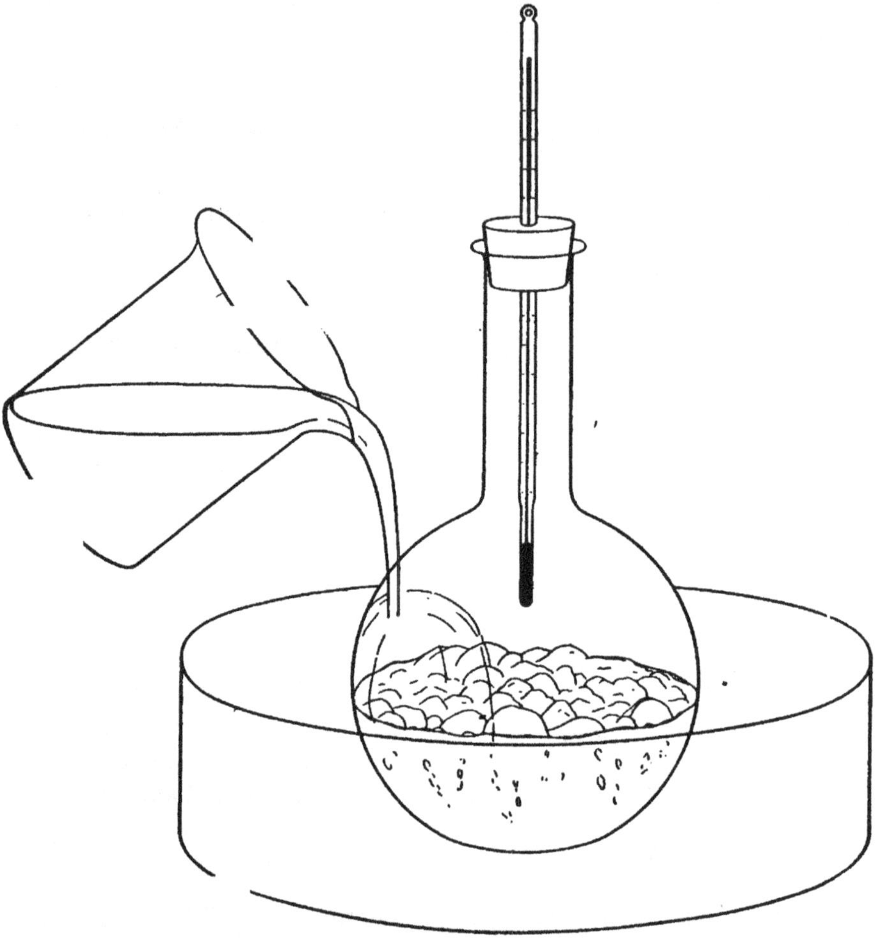

FIG. 24. Apparatus for showing the change in the boiling point of water with change of pressure

and pour some cold water over it. Try this a number of times, allowing the water to become cooler each time. After it has become quite cool remove the stopper from the flask.

By "ordinary conditions" we mean at atmospheric pressure, and it is under this pressure that water usually boils in an open dish. The water vapor which rose from the boiling water drove out the air from the flask. Then when the flask was cooled the water vapor inside condensed to liquid water, and the pressure on the surface of the water became lower. Then the water boiled again until the pressure of the water vapor stopped it. On cooling again more water vapor condensed, and so on. When the stopper was taken out the water-vapor pressure inside the flask was so small that the pressure of the outside atmosphere on the cork was felt. Water has a definite boiling point corresponding to each pressure, and it boils at 100° C. only when the pressure is that of the atmosphere at the sea level. At the top of a high mountain, where the pressure of the atmosphere is much lower, water boils at so low a temperature that it is impossible to cook some kinds of food by boiling them.

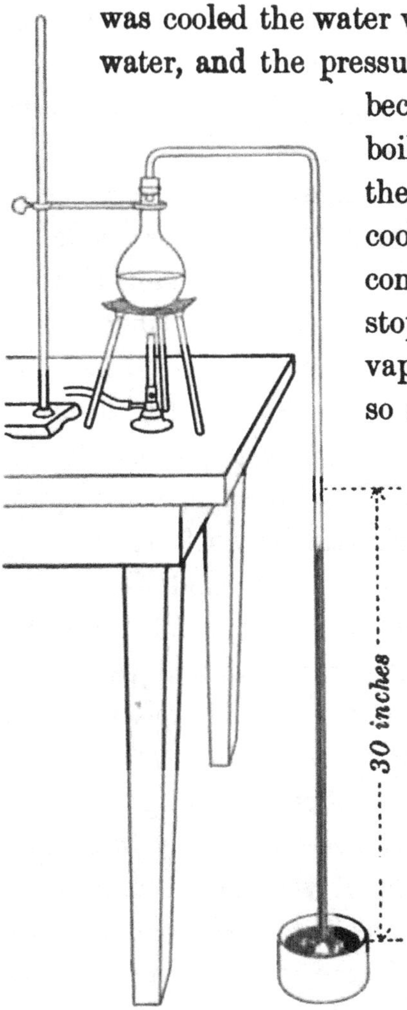

FIG. 25. Apparatus for measuring the vapor pressure at the boiling point

Experiment 38. *Demonstration.* Arrange an apparatus like the figure. The long arm of the glass tube should be about 90 cm. long. Sufficient mercury should be provided to nearly fill the long arm of the tube. Explain what takes place when the water is boiled and then allowed to cool.

47. Heat of vaporization. It takes about seven times as much heat to transform a gram of water into steam as to change a gram of ice into water, and the same amount of heat is given out when the steam changes back into water again. Steam is used to great advantage for heating purposes because its heat of vaporization is so great.

48. The water of the earth's surface. All the water of the earth passes through a remarkable series or cycle of changes. It is distilled by the action of the sun, that is, the sun furnishes us energy in the form of heat, which evaporates the water and lifts the water vapor so formed into the air. Here clouds form, and they are really very fine drops of water caused by the condensation of water vapor in the colder part of the atmosphere. This is very nearly pure water, and it falls as rain upon the earth's surface, sinks in, dissolving various substances on the way, and then flows from surface springs and brooks into rivers which carry the dissolved materials down into the ocean. All the plants and animals of the world are dependent for their life processes on water which is distilled in this way by the sun.

49. Water as a solvent. Beside this great importance of water in the economy of nature, it is also a very important substance in chemistry. Solutions of various substances are much more useful for many chemical purposes than the substances themselves, because solids do not as a rule react at all readily with one another. In order to make them react readily they must be brought into the liquid state, and this can be done either by melting them or by making solutions of them. The reason for this fact is evident when we remember that substances must, in order to react, come into contact with each other, and that solids can only come into contact over a very small surface, while liquids which mix have a very great surface of contact.

Water dissolves almost all substances to some extent. It seems in fact probable that all substances are soluble in water, although the amount dissolved may in many cases be so small as to be difficult of detection.

Experiment 39. Try the action of beet juice on a piece of glass. Grind the glass and beet juice together in a mortar.

QUESTIONS

1. Why does pure water not occur on the earth's surface?

2. Why is rain water not perfectly pure water?

3. Explain the different parts of the process called distillation.

4. At what temperature does water have the greatest density? Explain how this fact determines the way in which ponds and rivers freeze and thaw.

5. How do you know that ice is less dense than water?

6. Why do water pipes burst when they freeze? Has this fact any relation to the fact that water has its greatest density at 4° C.?

7. Describe the phenomena of supercooling and explain why it is that a body of water, cooled as that in the test tube was, can partially freeze.

8. What is meant by the heat of fusion of a substance?

9. Explain Experiment 37.

10. What is meant by the heat of vaporization of a substance?

11. Describe as fully as you can the cycle of changes through which the water of the earth's surface is continually passing.

12. What can you say of the importance of water for plant and animal life?

CHAPTER VII

HYDROGEN

50. Preparation. The amount of work which must be done to produce a metal from its oxide is a measure of the readiness with which that metal combines with oxygen (sects. 36–38). Potassium and sodium require a great deal of work to prepare them from their oxides, and they can be used to take the oxygen away from other substances on account of their great readiness to combine with it. They take away the oxygen to form an oxide with it, and leave the other constituent, which was previously combined with the oxygen, free to take some other form. Just as we used aluminium in Experiment 31 to prepare iron from one of its compounds (iron oxide), so we can use sodium to prepare another substance (*hydrogen*) from one of its compounds (water).

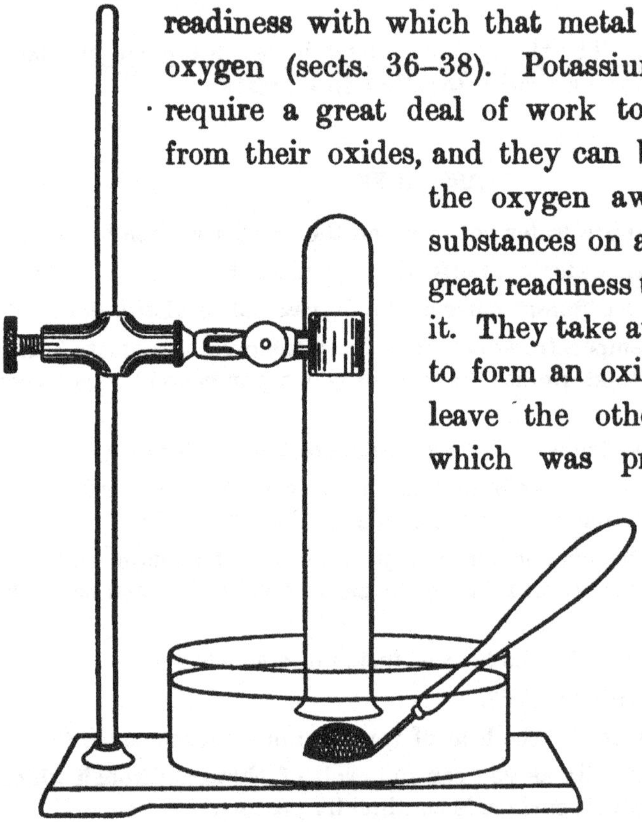

Fig. 26. Water decomposed by metallic sodium

Experiment 40. Place a small piece of sodium in a wire spoon (see Fig. 26). Hold the spoon under water and quickly invert a tube full of water over it. Repeat with small pieces of the metal until the tube is filled with gas.

Small quantities of sodium should be used, because the metal often contains impurities which cause slight explosions, and the larger the quantity used the more violent the explosions.

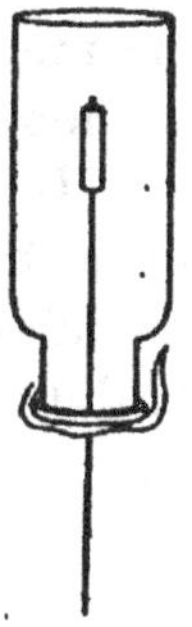

FIG. 27. Burning a jar of hydrogen

Experiment 41. Cover the mouth of the tube containing the gas. Lift it out of the water, uncover it, and quickly apply a flame to the mouth of the tube.

51. Properties. Hydrogen is the name of the gas which is produced in this way, and these two experiments have shown us a good deal about its properties. Its taste and odor we can examine later, but we know that it differs from oxygen, nitrogen, and carbon dioxide by being combustible, that is, combining rapidly with oxygen, with the production of heat and light.

Since nothing visible remains of the sodium, what facts can you state about the compound of sodium and oxygen which has been formed? Is it soluble or insoluble in water? Is its water solution colored?

This method of making hydrogen by means of metallic sodium is a rather slow and dangerous one, but there are many compounds beside water which contain hydrogen and many other metals beside sodium which are able to combine with the other constituents of these substances, leaving the hydrogen free to escape as a gas. Hydrochloric acid contains hydrogen and gives it up quite readily. Metallic zinc can combine with the other constituent of the hydrochloric acid, setting the hydrogen free.

Experiment 42. Prepare apparatus like that in Figure 28. Some granulated zinc is to be placed in a strong flask fitted with a rubber stopper with two holes, one for a delivery tube and the other for a dropping funnel. Add dilute hydrochloric acid from the funnel and collect small tubes full of the gas which escapes from the delivery tube over the pneumatic trough. Test each tube of gas by applying a flame to the mouth. A tubeful collected immediately after the action begins has all the properties of air. After the action has gone on for a time the gas in the tube ignites with a slight explosion. It is a solution of hydrogen and air. After the reaction has been going on still longer the gas ignites and burns quietly, just as the hydrogen prepared from sodium did.

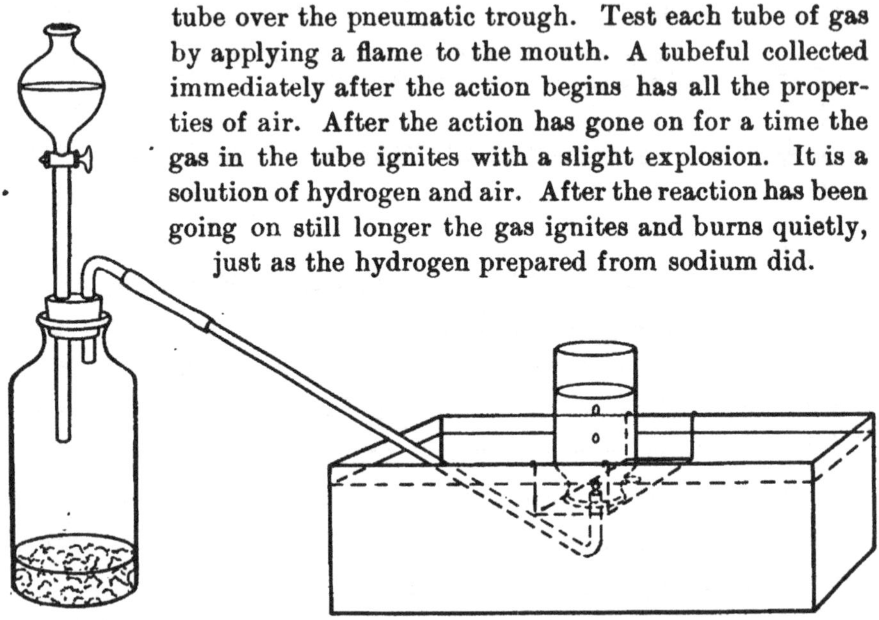

Fig. 28. Apparatus for generating hydrogen and collecting it in bottles

Caution! A mixture of hydrogen and air is explosive, and an explosion started in it will run through a long tube and into the generating flask, causing it to burst. Only small amounts of such a solution should be brought near a flame, and in every experiment where a jet of hydrogen is to be ignited, or there is a chance of an explosion running back into a closed flask, great care should be taken to keep flames away until only pure hydrogen is left in the generator.

52. Further properties. The properties of hydrogen can now be examined. It is a very light gas, only about one fourteenth as dense as air. It is colorless, tasteless, and odorless, and when it combines with oxygen, water is formed.

Experiment 43. Replace the delivery tube in the apparatus used in the last experiment by a glass tube which has been drawn out to a fine tip. Make sure that all air has been expelled from the flask. This is best done by performing the experiment after the hydrogen apparatus has been working for some time, and when a tubeful of

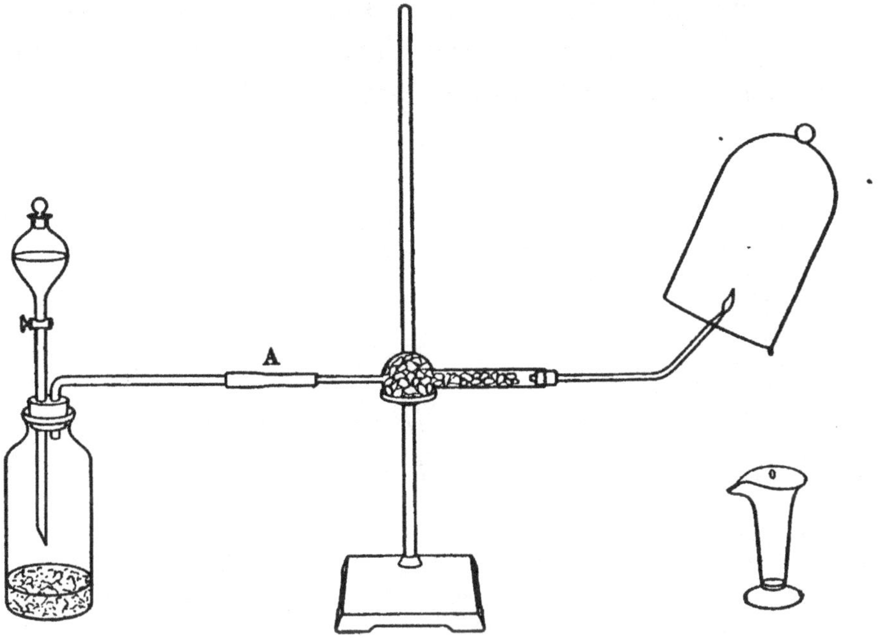

FIG. 29. Burning dry hydrogen in air

The drying tube should be inserted to prevent moisture from passing through from the hydrogen generator

gas collected over the trough burns quietly when lighted. Then replace the delivery tube by the tip at the rubber joint marked *A* in the figure.

Hold a cold glass beaker inverted over the flame. What forms on the glass?

Experiment 44. Place some mercuric oxide in a glass tube shaped like the one in Figure 30. Attach it to the hydrogen apparatus at the rubber joint. Heat the mercuric oxide carefully.

53. Reduction. Not only does hydrogen combine with free oxygen, but it can also take away the oxygen from compounds. Most of the compounds of oxygen with the heavy metals act as mercuric oxide does when heated in a current of hydrogen. The hydrogen takes the oxygen away from the metal and forms water with it, leaving the metal behind. Such a change from an oxide to a metal is called *reduction.* Hydrogen is a *reducing agent.* The change in the opposite direction, from a metal into its oxide, is called *oxidation.*

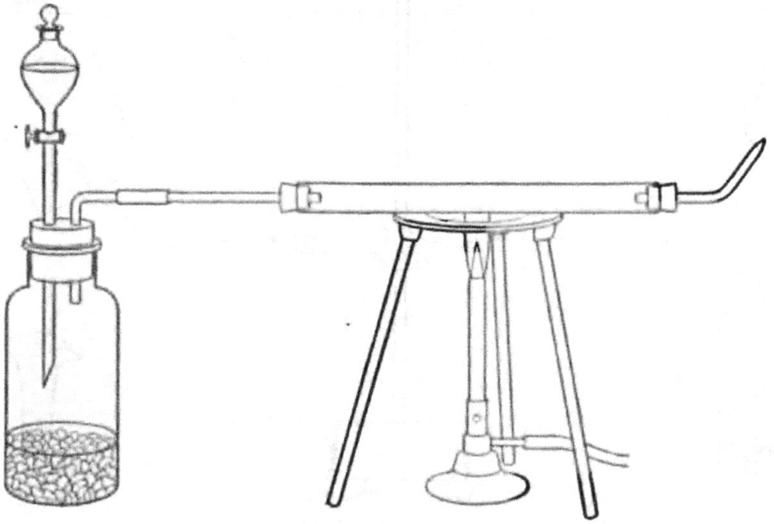

FIG. 30. Apparatus for the reduction of mercuric oxide by dry hydrogen

Experiment 45. Repeat the previous experiment, using black oxide of copper in place of mercuric oxide. After heating for a time shake out the powder remaining in the tube and rub it in a mortar. What is it? What became of the oxygen?

Try the same experiment with oxide of lead.

Try the same experiment with oxide of iron.

Experiment 46. Arrange an apparatus like that in Figure 31 with a boiling flask about half full of water connected by a glass and a rubber tube to a hard-glass tube containing finely powdered metallic iron. (The iron produced in the previous experiment will answer if the hydrogen has acted long enough to give complete reduction.)

Heat the iron in the tube by means of a burner and collect the gas which issues from the further end of the tube over a pneumatic trough. Test its properties. What is it?

54. Oxidizing agents. In Experiments 44 and 45 hydrogen acted as a reducing agent and reduced oxide of iron to metallic iron by taking away the oxygen with which it was

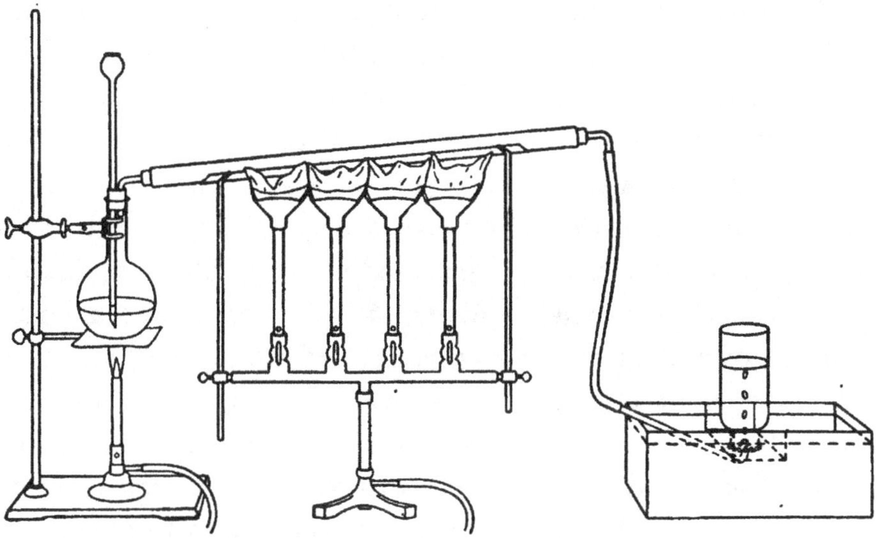

Fig. 31. Making hydrogen from water vapor and metallic iron

combined; and in Experiment 46 water vapor acted as an *oxidizing agent,* oxidizing metallic iron to iron oxide, while the hydrogen passed on as gas.

The reaction

$$\text{iron} + \text{water vapor} \rightleftharpoons \text{iron oxide} + \text{hydrogen}$$

can evidently be made to go in either direction, — from left to right if we pass water vapor over iron, and from right to left if we pass hydrogen over iron oxide, — so we have used the sign with two arrows in place of the ordinary equality sign. Exact experiments have shown that if we mix together iron, iron oxide, water vapor, and hydrogen in a closed vessel kept at a constant temperature, the reaction will go on in one direction or the other, depending on what is called the *concentration* of the water vapor and the hydrogen, and that the amounts of solid iron and iron oxide which we put into the vessel have nothing to do with the final result.

55. Concentration. *Concentration* means *the amount of a substance in unit volume,* and in Experiment 45 we arranged to have the concentration of the water vapor very small by sending a current of hydrogen over the iron oxide. This current of hydrogen carried away the water vapor as fast as it was formed, so that if we had taken out any quantity of the gas from the tube and analyzed it, we would have found it to consist of a small amount of water vapor and a large amount of hydrogen. In Experiment 46 we reversed this and sent a current of water vapor over the iron, which carried away the hydrogen, so that analysis of the gas in the tube in this case would have shown it to consist of a large amount of water vapor per cubic centimeter and a small amount of hydrogen.

If our four substances were mixed in a closed tube and kept at a constant temperature with, say, a large excess of hydrogen, then the reaction would go in the direction

$$\text{iron oxide} + \text{hydrogen} \longrightarrow \text{water vapor} + \text{iron};$$

but as soon as any amount of water vapor had formed it would begin to react with the iron in the direction

$$\text{water vapor} + \text{iron} \longrightarrow \text{iron oxide} + \text{hydrogen},$$

so that presently these two reactions would be going on equally in opposite directions and the reaction would come to a standstill.

The tube would then contain a mixture of hydrogen and water vapor which no longer changed, and which corresponded to the temperature at which the tube was kept. If the temperature is changed, the reaction goes on in one direction or the other until a new state of rest, corresponding to the new temperature, is reached.

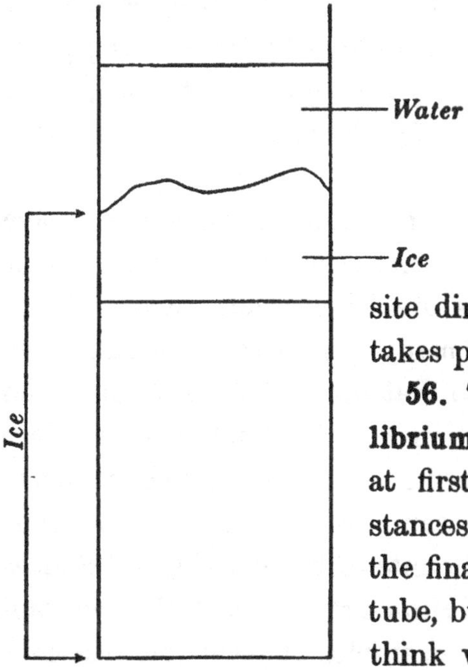

FIG. 32. Ice and water at the freezing point

Such a state of rest, in which two reactions are going on equally in opposite directions, so that no change takes place, is called *equilibrium.*

56. The effect of solids on equilibrium. It is not so easy to see at first just why the solid substances have nothing to do with the final condition of things in the tube, but it becomes evident if we think what will take place if we simply shut off part of the tube from the rest. We can shut off almost all of the gases from the solids and the amount of each gas per unit volume will not be changed. We could put a gram of iron and a gram of iron oxide into a tube containing any volume whatever of hydrogen and water vapor mixed in the right proportions for equilibrium and no change would take place, or we could put a hundred grams of one and a gram of the other into the tube with the same result. In the same way when equilibrium has once been reached in our tube, we can take away all but a very little of the iron and the iron oxide without changing anything.

This reminds one of the facts about ice and water. They can exist together at the freezing point, as we have found, and we do not need to inquire *how much* ice is present, or how much water. Figure 32 indicates why the freezing point is independent of the amount of ice which has formed, and a similar diagram can be drawn for the equilibrium of Experiment 46.

57. Reversible reactions. A reaction like the one we have been studying is said to be *reversible,* for very evident reasons. Not all chemical reactions are so evidently reversible as this one, and we shall find many which seem to run quite to an end in one direction so that none of the substances on the other side of the equation is left. Such reactions are said to be irreversible. It seems to be true that all chemical reactions are reversible if we can find accurate enough means of analysis, but often the amounts of the reacting substances which correspond to equilibrium are so small that we cannot easily find them. Other examples of reversible reactions which can be easily followed are to be found in sections 162 and 232.

QUESTIONS

1. How may hydrogen be prepared by the action of metallic sodium on water? How does the sodium act?

2. How may hydrogen be prepared from zinc and hydrochloric acid?

3. What are the physical properties of hydrogen gas?

4. What is the principal natural compound of hydrogen?

5. What is a reducing agent?

6. Describe an experiment which shows the reducing power of hydrogen.

7. What conclusions do you draw from Experiments 45 and 46?

8. Explain how it is possible for a reaction to go in either direction.

9. What is meant by equilibrium?

10. What is meant by the concentration of a substance?

11. Explain why the amounts of the solids, iron and iron oxide, which are present in the tube with hydrogen and water vapor, have nothing to do with the final state of equilibrium.

CHAPTER VIII

AIR

58. Nitrogen. We found in Experiment 25 that air contains about 20% of oxygen and that the rest of it is made up of another gas. We already know a good deal about this other gas, for we found that it is colorless and odorless and that it does not support combustion. In Experiment 18 the candle went out although there was plenty of the gas present. It is called *nitrogen*, and nearly four fifths of the atmosphere is made up of this gas.

Experiment 47. Fasten a piece of phosphorus to a bent wire and place it in a tubeful of air over water. Allow it to stand for an hour or more and then test the remaining gas with a glowing splinter or a match.

Nitrogen does not combine readily with other elements. It takes, in fact, a good deal of work to make it combine with anything, and this explains the large quantity of nitrogen which is free in the atmosphere. This fact also explains the cost of nitrogen compounds. As was mentioned in the case of magnesium, aluminium, and the alkali metals, they are expensive because the elements themselves can be made from their compounds only by the expenditure of a great deal of work. The reverse is true of nitrogen, for its compounds can only be formed from the element by the expenditure of work.

59. Combined nitrogen. Although nitrogen is such an inert substance it is a very necessary one for all plant and animal life. All living substances contain nitrogen in the combined state. Most of them are not arranged to cause nitrogen to enter into combination, and so they must get their supply from some other source. Some plants are able, with the help of bacteria which grow about and in their roots, to make atmospheric nitrogen combine with other elements, and these plants can therefore make their own supply of combined nitrogen. Many other plants, and animals as well, must have their nitrogen furnished to them already combined. One of the necessary constituents of a fertilizer is nitrogen in a form which can be easily used by plants, and animals all live on living food which contains a supply of the needed combined nitrogen.

Since work is required to make nitrogen enter into combination, work is given out when nitrogen compounds break up. Many explosives are examples of this; they contain nitrogen compounds and make use of the work which is given out when they break up.

Very many nitrogen compounds have bad odors. The smell of fertilizers, of burning wool and feathers, and many other bad odors are familiar examples.

60. Air is a solution. Although the composition of the air is very nearly constant, that is, there is almost exactly the same proportion of oxygen and nitrogen in a sample of air taken from any part of the atmosphere, air is not a compound, but a solution of these two gases in each other. There are several ways of showing this fact. One is to saturate water with air and then to pump out or boil out the air and examine the proportion in which the two gases are dissolved.

SIR WILLIAM RAMSEY (1852-   )

Discoverer of the rare gases of the atmosphere,—helium, argon, neon, krypton, and xenon

The gas which is pumped out of water in this way is no longer one fifth oxygen and four fifths nitrogen, but it is about one third oxygen and two thirds nitrogen. Each gas dissolves in the water according to its own solubility. If air were a compound it would have the same composition, whether taken from the atmosphere or pumped from solution in water.

61. Other constituents of the atmosphere. Besides nitrogen and oxygen, which make up nearly the whole of the atmosphere, several other gases are always found in air, but they are present in very small amount. Air always contains water vapor and a gas called *carbon dioxide*, and these are present in varying proportion, the amount of water vapor changing with the temperature and with climatic conditions. The variations in the amount of carbon dioxide we shall study about later (sect. 68).

Air also contains a number of rare gases which have been discovered within the past ten or fifteen years. One of these is *argon* and it is present in amount nearly equal to 1% of the nitrogen of the atmosphere. This gas is even more inert than nitrogen and does not form any compounds at all so far as we can find out, even though mixtures of argon and other elements are heated to the highest attainable temperature. The other rare gases of the atmosphere — *neon*, *krypton*, and *xenon* — are like argon in their properties.

Air can be liquefied by cooling it to a very low temperature and at the same time subjecting it to a great pressure. The very low temperature required is obtained by allowing a portion of the compressed air to expand, for the expansion of a gas always means the absorption of heat.

Liquid air is now made commercially on a large scale. The nitrogen and oxygen are then separated by a process of fractional condensation and sold in a nearly pure state. Oxygen made in this way is used for welding flames and in other processes where a very high temperature is necessary.

QUESTIONS

1. What are the properties of nitrogen?

2. How did we prepare nitrogen?

3. Why is it that compounds of nitrogen have a market value while the element itself costs nothing at all?

4. How can you show that air is a mixture?

5. When nitrogen and oxygen are mixed in the proportion in which they exist in air, no change in temperature takes place. What does this suggest?

CHAPTER IX

CARBON

62. Forms of carbon. One of the most important elements in nature, and one on which all plant and animal life is directly dependent, is *carbon*. This element occurs in several forms, lampblack, graphite, and diamond all being very

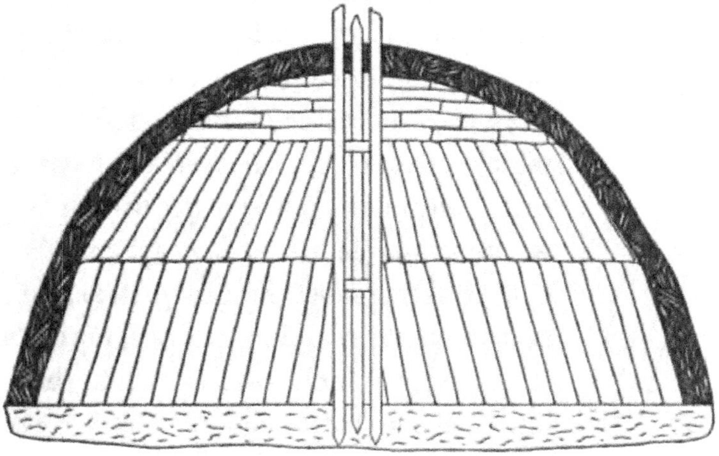

FIG. 33. A primitive form of charcoal furnace, in which wood is changed into charcoal, — nearly pure carbon

nearly pure carbon, while ordinary coal, anthracite, peat, and many other similar substances contain a very large percentage of carbon. These have all been formed from plants which have changed into these forms by slow processes very much like the one which results in the production of charcoal from wood.

63. The value of coal. Carbon in the form of coal is the source of nearly all the energy which is used on the earth for technical purposes. The other sources of power — water, winds, and tides — furnish only a very small percentage of the energy needed for running our factories and keeping us warm in winter. Carbon alone is not sufficient as a supply of energy. In order to get work out of it, it must be burned, and in order to burn it a supply of oxygen is necessary. But carbon is a solid which occurs only in special places on the earth, while oxygen is a gas which is everywhere available in the atmosphere. So while one of these substances is just as necessary as the other in order that we may set free the energy which results when they combine, carbon costs money while oxygen is free to all.

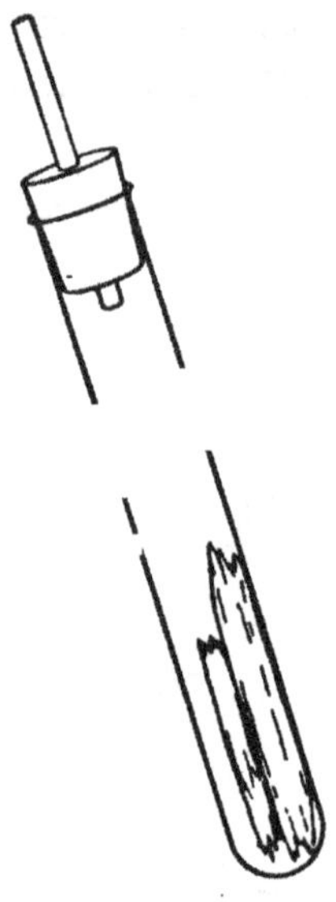

FIG. 34. Distillation of wood to obtain charcoal

It is the energy contained in the carbon for which we pay when we buy a ton of coal. The products of combustion are sent up the chimney as rapidly as possible and the ashes are thrown away.

64. Foods. All foods contain carbon, and it is at the expense of the energy of our food and the oxygen which we breathe that we carry on all our bodily and mental processes. A part of this energy we use as heat, for our bodies have a temperature above that of the surrounding air most of the time. Another part is used up in the form of work when any bodily exertion is performed, and other parts are used up in carrying on the various processes which go on within the body.

Experiment 48. Heat a piece of charcoal in a hard-glass tube arranged as in Figure 35. Pass air over it from the gas holder and dip the mouth of the delivery tube into limewater.

When carbon is burned in oxygen the product is carbon dioxide (compare with Exp. 18). Most organic substances — wood, coal, etc. — yield other substances besides carbon dioxide when they burn.

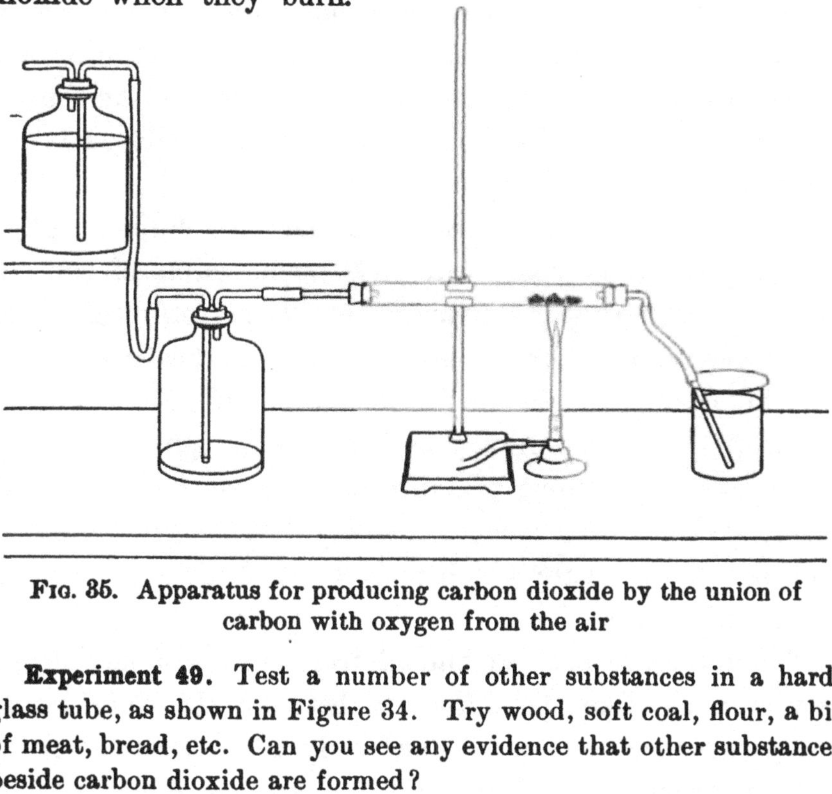

FIG. 35. Apparatus for producing carbon dioxide by the union of carbon with oxygen from the air

Experiment 49. Test a number of other substances in a hard-glass tube, as shown in Figure 34. Try wood, soft coal, flour, a bit of meat, bread, etc. Can you see any evidence that other substances beside carbon dioxide are formed?

65. Allotropic forms. In order to find out whether a substance is *pure* carbon or not, it must be burned in oxygen and the weight of the carbon dioxide which is formed must be determined. If the carbon in the carbon dioxide has the same weight as the substance used in the experiment, then the latter was pure carbon. Analyses like this show that pure charcoal, graphite, and diamond all give the same weight of carbon dioxide for the same weight of substance burned (they all give 44 parts of carbon dioxide for 12 parts of substance taken), and the weight of carbon in the carbon dioxide formed is the same as the weight of the whole substance burned. While the physical properties of these three substances are very different, they have the same chemical properties. This means that when charcoal, graphite, and diamond enter chemical reactions, they give the same products in the same proportions, independent of the nature of the process in which they are taking part.

When an element has several forms with different physical properties, these are called *allotropic* forms of the element. It is possible to change these forms into each other. Charcoal can be changed into graphite by heating it to a very high temperature, but in order to produce diamond from either of the other forms of carbon a very high temperature and also an enormous pressure are necessary. Very small diamonds have been made artificially, but the great pressure and high temperature which must be used make the process a very difficult one.

The difference between the allotropic forms of an element is very much like the difference between the solid, liquid, and gaseous states of a substance. In both cases the chemical properties of the different forms are the same as far as the formation of compounds is concerned. But the physical properties of the different forms are very different, and there is also a difference in the amount of energy they contain. Just as energy is liberated when water changes into ice or steam into water, so energy is set free when charcoal changes into graphite or diamond.

66. Carbon monoxide. When carbon burns incompletely, that is, without a sufficient supply of oxygen to form carbon dioxide, another compound of carbon and oxygen is produced which is called *carbon monoxide.* This substance is a gas, colorless, odorless, and very poisonous. In countries where open pans of charcoal are used for heating, death from carbon monoxide poisoning sometimes occurs. This gas is one of the most treacherous and dangerous of the various gases

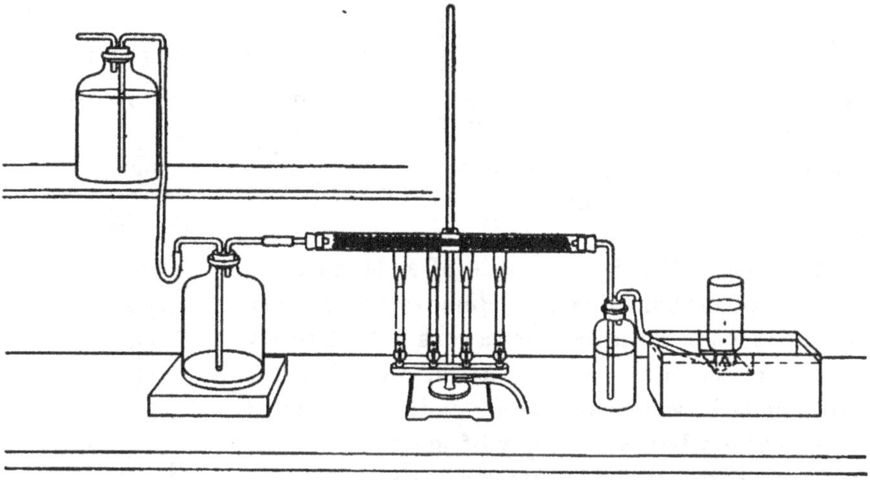

FIG. 36. Apparatus for making carbon monoxide by the union of carbon with oxygen

which form in coal mines as the result of explosions and of slow-burning fires in the coal. It results from the incomplete combustion of carbon, and it can combine with more oxygen to form carbon dioxide. It is therefore a combustible gas, and its blue flame can always be seen flickering over a coke fire.

Experiment 50. *Demonstration.* Arrange apparatus as in Figure 36 and send a slow stream of oxygen through a hard-glass tube filled with small pieces of charcoal heated to a bright red heat. Collect the gas which passes over in a pneumatic trough and test its properties.

67. Carbon dioxide. In Experiment 17 we learned that carbon dioxide does not support combustion, and we used a reagent, limewater, as a test for this gas. We also found by that experiment that this gas is heavier than air, since it stayed in the open flask and did not rise and escape. It is so much heavier than air that it can be poured from vessel to vessel like a liquid.

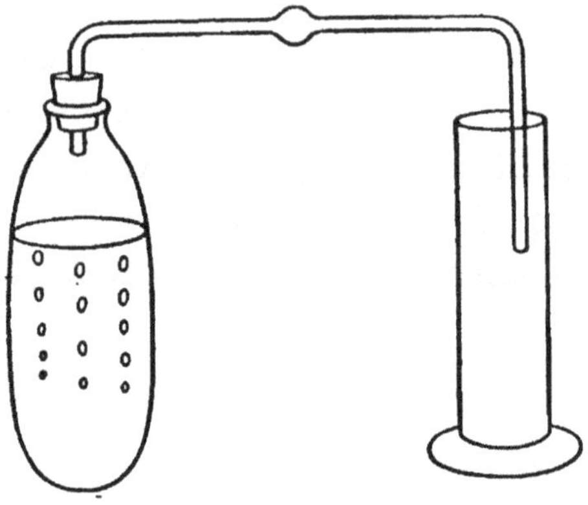

FIG. 37. Obtaining carbon dioxide from soda water

Experiment 51. Use a bottle of soda water as a source of carbon dioxide. Remove the stopper and quickly insert in its place a cork with delivery tube. Collect the gas in bottles, warming the bottle of soda water by placing it in a beaker of warm water when the supply of gas begins to diminish. Test the gas for its properties.

Experiment 52. Arrange an apparatus as the Figure 38, partly filling the generating flask with broken pieces of marble. The delivery tube should reach nearly to the bottom of the wide-mouthed receiving flask. Pour dilute hydrochloric acid over the marble and collect the gas by allowing it to fill the bottles and force the air out. See when a bottle is full by applying a lighted match to the mouth. Make all the tests you can think of on this gas. Lower a candle into a bottle of the gas. Pour a little limewater into another bottle and shake it. Lower a short piece of lighted candle into a wide-mouthed beaker and pour carbon dioxide from one of the bottles over the candle.

FIG. 38. Generating flask for carbon dioxide

Carbon dioxide is a substance of great importance in nature. It is always present in small quantities in the atmosphere, because all animals and plants breathe out carbon dioxide, which has been produced in their bodies by the combination of carbon with the oxygen they have breathed.

FIG. 39. Effect of carbon dioxide on combustion

Experiment 53. Leave some limewater standing in an open dish for a time. Is there evidence that there is carbon dioxide in the air of the room?

Experiment 54. Arrange an apparatus like the figure and draw air from the room through limewater. Try air from out of doors in the same way.

68. The carbon-oxygen-carbon dioxide cycle. When carbon combines with oxygen to form carbon dioxide a large amount of heat is set free. It is evident from this that energy

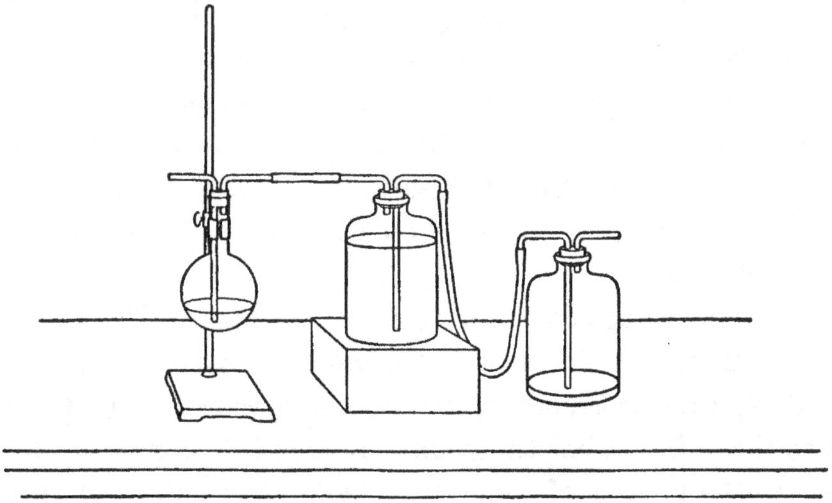

FIG. 40. Apparatus for showing the presence of carbon dioxide in air

must be supplied from some outside source if carbon is to be separated from oxygen, leaving the two elements in the free state. All plants and animals breathe in oxygen and breathe out carbon dioxide, and if there were no simple process in nature by means of which the carbon and oxygen could be separated, the atmosphere would grow constantly poorer in oxygen and constantly richer in carbon dioxide. There is in the world a remarkable cycle of changes in these substances accompanied by important energy changes.

All of the energy which is available for our use, with the exception of very small quantities, comes to us from the sun in the form of light, and is transformed in the green leaves of plants into a form which we can use. By the aid of this absorbed energy the green leaves are able to reverse the ordinary process of combustion. They split up the carbon dioxide gas, which they take from the atmosphere, into oxygen, which they breathe out, and carbon, which they use in the formation of the substances they need for their life processes. It is only the green plants which have this power; all other plants are dependent for their supply of energy on already formed carbon compounds, which they burn with oxygen to form carbon dioxide. Nor can animals use the light from the sun in this way. They are all dependent for their existence on the carbon formed by the green plants.

All living things carry on their life processes at the expense of the carbon which they contain and the oxygen which they breathe; this statement applies to the green plants as well as to the animals, but the plants are able to absorb light and reverse the process of combustion, with the production of an *excess* of carbon and oxygen.

It should be noticed that while the carbon-carbon dioxide-oxygen cycle is a closed one, that is, simply goes round and round without loss or gain of anything, the supply of energy from the sun is continually kept up. The life processes on the earth can only go on while this supply is kept up, and if it should cease, the plants could no longer reverse the process of combustion, and the present conditions of life on the earth would come to an end.

Experiment 55. Place some green, growing plants in water with a funnel full of water inverted over them. Put this apparatus in a bright light, and after a little gas has collected in the neck of the funnel collect it over the pneumatic trough and examine its properties. What is it?

Remove all the gas from the funnel and allow the plants to stand in the dark over night. Test the water for dissolved carbon dioxide with limewater.

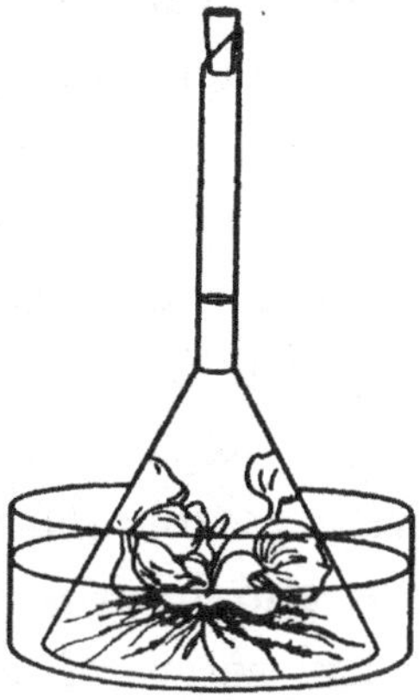

Fig. 41. Apparatus for collecting the gases given off by plants

69. Carbon compounds. Plants do not make elementary carbon from the carbon dioxide of the air, but they make carbon compounds. Most of the carbon compounds which are formed in nature contain hydrogen also. Besides carbon, ordinary coal contains hydrogen, nitrogen, and sulphur. Crude petroleum is a solution of a very great number of compounds of carbon and hydrogen, which are so similar to one another that it is difficult to separate them. When crude petroleum is distilled, the compounds which have the lowest boiling point pass over first and can be collected. This first product of distillation is called *benzine*, *gasoline*, or *naphtha*. A mixture of the vapor of these compounds with air is explosive, and they are therefore used for driving motors. They are also very good solvents for many substances which are not dissolved by water, for instance, fats and oils. As the distillation of the crude petroleum goes on, its boiling point rises, and at a higher temperature the less explosive oils are produced, which are collected separately and sold under the name of *kerosene*. At still higher temperatures heavy lubricating oils come over, and then still thicker. substances like vaseline. Finally, at still higher temperatures, hard, waxy substances separate, and these form the substance *paraffin*.

70. Hydrocarbons. Each of the substances in this series is a solution of a number of carbon compounds, and the whole number of such compounds which is contained in the crude petroleum is very large. They are all composed of carbon and hydrogen, and they are called *hydrocarbons*. The entire series of compounds is called the paraffin series of hydrocarbons. There are other series of hydrocarbons which can be produced by the distillation of other naturally occurring organic substances. One of these series consists of substances which are like *benzol*. A whole set of compounds, of which naphthalene is a type, can be made from coal tar. Moth balls are made of impure naphthalene.

Each of these hydrocarbons of the paraffin, benzol, and naphthalene series can form a great number of compounds which are like each other and which have properties varying in a definite way with the boiling point of the compound. The study of these hydrocarbons and the many other substances which can be derived from them makes up what is called organic chemistry, which is, in fact, the chemistry of the compounds of carbon. In organic chemistry even more than in inorganic chemistry the study of the general laws is of the greatest aid, for compounds which belong to one series have properties which vary in a definite way with their physical properties.

71. The chemical substances of the animal body. The substances produced in the bodies of animals are very complex ones indeed, but some simple general reactions are fundamental for the processes of life. We know that all living things depend on some source of energy to carry out the processes which are necessary for life. The most usual, and, in fact, almost the invariable, source of this energy is the combustion of carbon with the oxygen of the air. In such complicated organisms as the human body the oxygen is carried to the parts where it is needed by some special arrangement like the blood. The blood of animals contains substances which take up oxygen very easily and give it up again almost as easily.

72. Hæmoglobin. This name is given to a substance contained in the red blood corpuscles, which is oxidized in the lungs, by contact with air, to oxy-hæmoglobin. This is then forced by the action of the heart to all parts of the body, giving up its oxygen where it is needed. The product of combustion — carbon dioxide — returns to the lungs through the veins and is removed when the breath is exhaled.

Experiment 56. Blow the breath through a tube into limewater.

73. Food. Food furnishes the carbon for this combustion, and only a part of what we eat is burned directly to carbon dioxide. Other parts are used to build up the tissues of the body, especially when one is growing fast. When hard work is being done food enough must be consumed to supply energy for the external work, and a correspondingly large amount of oxygen must be taken into the lungs to assist in the reaction.

These things are all cared for automatically by regulating devices in the body, and we do not have to trouble ourselves to remember to breathe faster when we are walking fast or running, nor do we have to measure the amount of work we have done in order to decide how much food we should eat. The regulating devices of the body force us to breathe rapidly when we are performing severe muscular work, and our appetite is attended to in the same way.

QUESTIONS

1. In what forms does the element carbon appear in nature?

2. What is coal?

3. What other substance is always just as necessary as the coal? Explain the difference in the cost of these two substances.

4. Is coal valuable because of its weight? Explain.

5. What uses do we make of the energy developed in our bodies from the food we eat and the air we breathe?

6. Beside carbon dioxide what other substance results from the combustion of most organic substances?

7. How would you determine whether a substance was pure carbon or not?

8. What is meant by the word "allotropic"?

9. What are the physical properties of carbon monoxide? How can it be made? What are its chemical properties?

10. What are the physical properties of carbon dioxide? What are its chemical properties?

11. Explain the part which carbon dioxide takes in plant and animal life.

12. State what you know about the hydrocarbons of the paraffin series.

13. Tell what you can of the cycle of changes which takes place between carbon, oxygen, and carbon dioxide in the human body.

CHAPTER X

ACIDS AND BASES

74. Salt. One of the commonest substances which we usually think of as a "chemical," and almost the only pure substance which we use in food, is common salt. This substance contains the element sodium and another element which we shall soon study about, which is called *chlorine.* The chemical name for common salt is *sodium chloride.* All salts, in the broad use

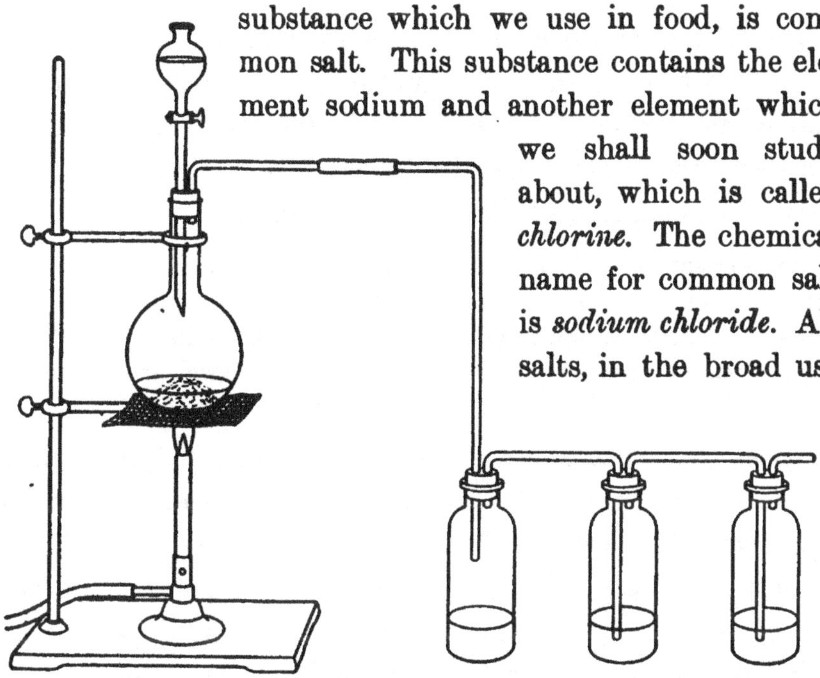

FIG. 42. Apparatus for making hydrochloric acid

of this word, are substances which contain a metal and another constituent.

Experiment 57. Arrange an apparatus like that in Figure 42. Place some salt in the flask and attach the delivery tube to a series of three absorption bottles. The end of the tube should not dip below the surface of the water in the first bottle. Pour concentrated sulphuric acid on the salt, and heat. Test the liquid in the bottles with litmus paper. Dilute some of the liquid with a good deal of water and taste it. Test with litmus paper the gas which issues from the mouth of the delivery tube.

75. The tests for an acid. The change of color from blue to red is a test for an *acid.* Another very delicate test for an acid is its taste, and a very small amount of an acid in a large amount of water can be detected in this way. Another test for an acid is its behavior toward metals.

Experiment 58. Place some powdered magnesium in a test tube with delivery tube. Add a little hydrochloric acid. Test the properties of the gas which is produced. Try in the same way all the other acids at hand, including weak ones like vinegar and lemon juice.

These three properties — a sour taste, the turning of blue litmus paper red, and the production of hydrogen by reaction with a metal — are specific properties of a large class of substances which are called acids. The most important and certain of these properties is the formation of hydrogen by the action of a metal. All acids contain hydrogen which can be set free in this way, and while very many other substances contain hydrogen, they differ from the acids in not having this important property. There are a large number of acids, but we shall study about only a few for the present. The commoner ones are (beside hydrochloric acid) sulphuric acid, nitric acid, acetic acid, which gives vinegar its sour taste, and citric acid, which is one of the acids present in lemons. The first four of these we shall have occasion to use quite often.

76. Bases. There is another class of compounds which are in many respects the opposite of the acids. They are called *bases,* and we have already used one of them in

Experiment 24 to combine with the water and carbon dioxide formed during the combustion of the candle. This particular one is called caustic soda, or *sodium hydroxide.*

Experiment 59. Make a solution of caustic soda in water and test it with litmus paper which has been made red by an acid. Dilute some of the solution with a good deal of water and taste it. Try the feeling of the solution between your fingers.

77. Tests for a base. Bases in solution all have the property of turning red litmus paper blue, and they also have the peculiar taste which was experienced and the slippery feeling as well.

Experiment 60. Add a few drops of litmus solution to some water in a glass. Add first a drop or two of hydrochloric acid and then caustic soda solution drop by drop.

A solution which turns red litmus blue is called an *alkaline* solution.

Experiment 61. Make a solution of about 10 grams of sodium hydroxide in 100 grams of water. Put part of this into a beaker, add a drop or two of litmus solution, and then add hydrochloric acid until the blue color just changes to red. If a little too much hydrochloric acid is added, a drop or two of the alkaline solution will bring it back into such a condition that a drop of acid will make it red or a drop of caustic soda solution will make it blue. The solution is now *neutral.* Evaporate the solution in an evaporating dish, first over a free flame and then on a water bath or a sand bath, until it is nearly dry. What separates out? Test its properties. Taste it.

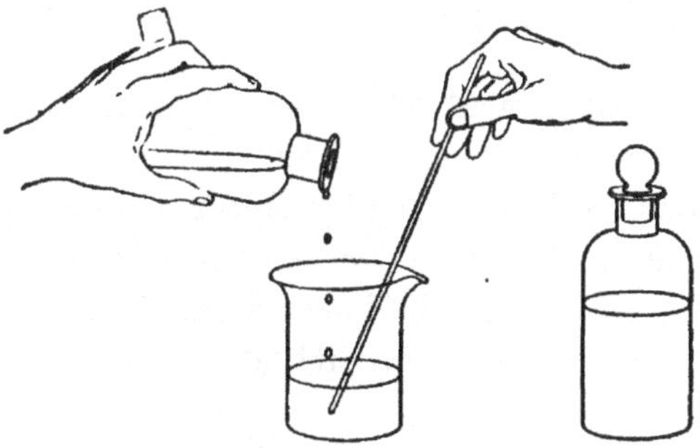

FIG. 43. The neutralization of an acid by a base

78. Neutralization. From hydrochloric acid and a base, sodium hydroxide, we have made a salt—common salt, which has the chemical name of sodium chloride. The sour acid, which attacks metals and behaves in general like a very strong chemical, and the strongly alkaline caustic soda have both disappeared; the resulting compound, common salt, has none of the chemical or physical properties of the two substances from which it was made.

There are many bases beside caustic soda, and each of them can combine with any acid to form a salt. If we have in the laboratory ten acids and ten bases, we can make a hundred different salts from these twenty substances, and as there are hundreds of acids and bases it is evident what a great number of salts there must be. Most of the precipitates which a chemist uses as tests for various substances, because of their special properties, are salts. The white precipitate which forms when limewater and carbon dioxide are brought together is a salt. Limewater is a base and a solution of carbon dioxide in water is an acid.

Experiment 62. Test limewater with a piece of red litmus paper. A piece of prepared red paper should be used, and not one which has been made red by dipping it into acid, for the paper in the latter case will have so much free acid adhering to it that there will not be enough of the base in the dilute solution which lime makes with water to combine with all of it.

From Experiment 61 it is evident that a perfectly definite amount of a base combines with a perfectly definite amount of an acid to form a salt. Experiment 62 has shown that limewater is a base. Now we will examine the acid properties of *carbonic acid*, which is produced when carbon dioxide dissolves in water. Carbon dioxide itself cannot, of course, be an acid according to our definition, for it contains only carbon and oxygen and no hydrogen.

Experiment 63. Pass some carbon dioxide gas, produced by the action of hydrochloric acid on marble, into water containing a few drops of litmus solution.

Carbonic acid is a weak acid, and this is the reason why the color produced with litmus is not a bright red.

Experiment 64. Prepare the sodium, potassium, and calcium salts of hydrochloric, nitric, and sulphuric acids and examine their properties.

79. Salts. The salts of various acids and bases have different physical and chemical properties. Some of them are very soluble in water, as calcium chloride, and a great many are soluble in about the same degree as common salt and potassium chloride. Others are very slightly soluble, as calcium sulphate. The greater part of all the chemicals which are to be found on the laboratory shelves are salts, and most of them are soluble in water. It has already been mentioned that substances in solution react much more easily and rapidly than they do in the solid state; we therefore keep a stock of soluble salts, which we can dissolve in water and so prepare different solutions which react as rapidly as possible.

80. Hydrogen ion. From our experiments with acids we may conclude that the hydrogen of acids does not behave as elements usually do. For example, a compound of carbon may be either a gas, a liquid, or a solid; it may be colorless or colored; there is no common property by which we can immediately recognize the fact that any substance is a compound of the element carbon. The same is true of compounds of oxygen or nitrogen. Neither do compounds of hydrogen show any common properties which would enable us to decide immediately that hydrogen is present, *except the ones which belong to the class of substances we have called acids.* These *all* give off hydrogen with metallic magnesium, have a sour taste, and turn blue litmus paper red.

Some elements may be immediately recognized in all of their compounds by characteristic properties without breaking up the compounds. When this is true they are are called *ions* and they are said to be in the *ionic state* when they show these properties. This is a peculiar state, just as diamond is a peculiar state of the element carbon.

Acids are substances which contain hydrogen ion, and to test for its presence it is only necessary to dissolve a substance in water and test it with blue litmus paper; if it turns red, hydrogen ion is present. Blue litmus paper is a reagent for hydrogen ion, just as limewater is a reagent for carbon dioxide. Metallic magnesium and a sour taste are also reagents for hydrogen ion, for every substance which, when dissolved in water, will turn blue litmus red will also taste sour and give off hydrogen gas with magnesium.

81. Chloride ion. Hydrogen is not the only element which can take on the ionic state. Many of the other elements, and especially the metals, can show a similar set of properties common to all their compounds when dissolved in water, and there are many compound substances which can act in the same way.

We know that hydrochloric acid contains hydrogen ion, since it gives all the tests for it. This acid contains another element, chlorine, which can take the ionic state, and in this condition it is called *chloride ion.* A solution of silver nitrate, which is made from silver and nitric acid, is a test for chloride ion, producing a white precipitate with every solution which contains it.

Experiment 65. Add silver nitrate solution to dilute hydrochloric acid.

Add a drop of hydrochloric acid to half a liter of water and test this dilute solution with silver nitrate solution.

This test is a very sensitive one, and the presence of a very minute quantity of chloride ion can be recognized by its use.

This same test for chloride ion will also show its presence in all salts made from hydrochloric acid. We know that common salt can be made from hydrochloric acid, and we have made other salts from it.

Experiment 66. Test all the salts you have made from hydrochloric acid for the presence of chloride ion.

Just as there are many hydrogen compounds which do not contain hydrogen ion, so there are also many chlorine compounds which do not contain chloride ion.

Experiment 67. Dissolve some potassium chlorate in water and test for chloride ion.

Potassium chlorate is a substance which contains chlorine, as has been shown by breaking it up and proving its presence, but the chlorine is not present in potassium chlorate in the form of chloride ion.

82. Sulphate ion. Sulphuric acid contains, beside hydrogen ion, another ion which is called *sulphate ion*, the name being taken from the fact that the salts of this acid are called sulphates. Barium chloride is a reagent for this ion.

Experiment 68. Add a few drops of sulphuric acid to half a liter of water and add to this a few drops of a solution of barium chloride.

Describe the difference in appearance between this precipitate and that produced when silver nitrate is added to a solution of sodium chloride.

The precipitate in one case is said to be "powdery," and in the other "flocculent."

Sulphate ion differs from hydrogen ion and chloride ion in containing more than one element. It is a compound ion, made up of sulphur and oxygen.

Experiment 69. Prepare various sulphates by neutralizing sulphuric acid with various bases. Dilute solutions should be used. Test for the presence of the sulphate ion with barium chloride solution.

83. Hydroxide ion. We have already found that bases have the common property of turning red litmus blue. This suggests the question whether all bases in solution contain a common ion. The answer is that they do contain such an ion, which is composed of oxygen and hydrogen and which is called *hydroxide ion.* It is therefore a compound ion like sulphate ion, and not an elementary one like hydrogen ion and chloride ion.

All bases are salts containing this ion combined with another ion, and the second ion is usually the ion of a metal. Caustic soda (or sodium hydroxide) solution contains sodium ion and hydroxide ion, and limewater contains calcium ion and hydroxide ion. Nearly all the metals can form elementary ions, and each of these ions can form a base with hydroxide ion.

84. Salt solutions. So far we have found ions only in solutions in water, and the question may be asked, Is it possible that ions should exist alone, in a pure state, as hydrogen, oxygen, and carbon do? Ions never do exist alone or in the pure state, but only in solutions; and no solution can ever contain one ion only, but if it contains ions at all there will always be at least two present at the same time.

In hydrochloric acid we found hydrogen ion and chloride ion, and sodium ion is always present in caustic soda along with hydroxide ion. This same sodium ion (for which there is no very evident or simple test) is present in common salt together with chloride ion. Every salt which is made from caustic soda by neutralizing it with any acid always contains sodium ion together with the other ion which was present in the acid with the hydrogen ion, and these salts are all called sodium salts. In just the same way all the salts which can be made from sulphuric acid contain sulphate ion and are called sulphates. The salt which is made from caustic soda and sulphuric acid contains both sodium ion and sulphate ion and is called sodium sulphate. The name for each salt consists of two parts which describe what ions it contains. The metallic ion is usually named first and the other ion second. Examples are sodium chloride, sodium sulphate, calcium chloride, calcium sulphate, etc.

85. Classes of ions. All ions can be arranged in two groups. In one group we place the ions which make acids when combined with hydrogen ion, — sulphate ion and chloride ion, for example. These are called acid ions or *anions*. In the other group we place the ions which form bases when combined with hydroxide ion, — sodium ion, for example. These are called basic ions or *kations*. It may be stated as a general law that only ions which belong to opposite groups can unite to form salts. Chloride ion does not unite with sulphate ion, nor does sodium ion unite with calcium ion. But every kation can unite with every anion to form a salt.

86. Precipitates. When the salt formed by the union of a kation with an anion is insoluble (or slightly soluble) in water, it will appear in the solid form every time its ions meet in solution. The precipitates by means of which we tested for chloride and sulphate ions are salts of this kind, which are only very slightly soluble in water. All silver-salt solutions contain elementary silver ion, and when silver ion meets with chloride ion in a solution they unite to form insoluble silver chloride. All barium salts contain barium ion, and whenever this meets with sulphate ion they combine to form insoluble barium sulphate.

Since two different ions are always concerned in the formation of an insoluble salt in solution, either of them will serve as a test for the other. So any solution which contains chloride ion will serve as a test for the presence of silver ion, and any solution containing sulphate ion will show the presence of barium ion by producing a precipitate of barium sulphate.

When the substance so formed is very slightly soluble, a very small amount of one ion can be detected in a large amount of solution.

Experiment 70. Test the tap water for the presence of the chloride ion and the sulphate ion. It may be necessary to let the test tubes stand for a time before the precipitates become clearly visible.

87. The formation of salts. We prepared common salt, sodium chloride, by neutralizing sodium hydroxide with hydrochloric acid. Sodium chloride contains sodium and chloride ions. What has become of the other ions, the hydroxide ion and hydrogen ion, which were in the base and the acid when we started the process? They seem to have disappeared, for all their properties have vanished.

They have united to form *water*. Like water, hydroxide ion consists of oxygen and hydrogen, but it contains less hydrogen than water, and it turns into water when it unites with more hydrogen. The reaction went on in solution, where there was already a great deal of water, so the formation of a little more could not be detected. If we had excluded water, the formation of water during the reaction could be shown.

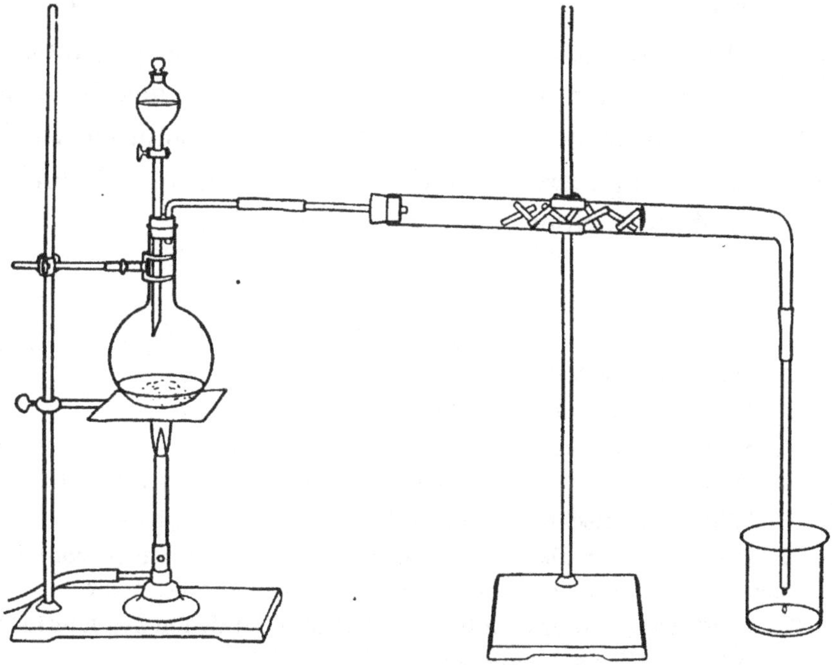

FIG. 44. The combination of hydrochloric acid with caustic soda

Experiment 71. Arrange apparatus as in the figure, with a piece of solid caustic soda in the hard-glass tube. Pass a current of hydrochloric acid gas through the tube from the generator described in Experiment 57, and heat the caustic soda gently with a burner.

Whenever an acid is neutralized by a base, a salt is formed and the hydrogen ion and hydroxide ion unite to form water, which is in a sense also a salt, but an exceptional kind because it does not contain a metal.

Beside this method of neutralization, salts can also be formed in other ways. Many metals have the property of replacing the hydrogen of an acid, driving it out of the ionic state to become hydrogen gas, while the metal leaves the elementary metallic state and changes into the ionic state. Magnesium does this so readily that we have used it as a reagent for hydrogen ion (Exp. 58). And in the preparation of hydrogen gas for our experiments (Exp. 42) metallic zinc was used in the same way. If we examine the solution which was left in the generator in this experiment, we will find a zinc salt of hydrochloric acid, — zinc chloride.

QUESTIONS

1. What are the tests for an acid?
2. What are the common acids?
3. What are the properties of a base in solution?
4. What is meant by the neutralization of an acid by a base?
5. What is formed when an acid is neutralized by a base?
6. How can you distinguish between a substance which contains hydrogen in combination and one which contains hydrogen ion?
7. How could you test for the presence of chloride ion in a solution?
8. What test for sulphate ion have you used?
9. What ion is peculiar to all bases in solution? Tell what you can of its properties and the tests you have used for its presence.
10. How are salts named?
11. How does the fact that a salt is slightly soluble enable you to detect either of its ions in a solution?
12. What methods have you used for the preparation of salts?
13. Tell what you can of the physical properties of the salts you are acquainted with (solubility, color, etc.).
14. When an acid in solution is neutralized by a base what beside a salt is formed? How can you prove this?

CHAPTER XI

COMBINING WEIGHTS

88. Quantitative neutralization. In Experiment 60 we found that a very sharp color change can be produced by adding acid and alkali to a solution of litmus. This sharp point of *neutrality*, as it is called, can be used to find the amount of an acid which will exactly neutralize a given amount of an alkali.

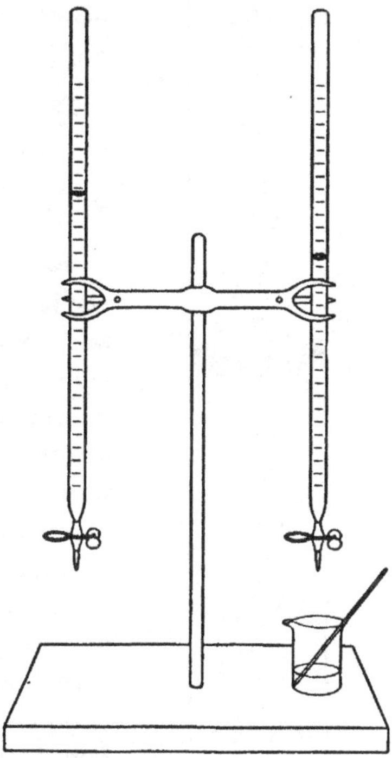

FIG. 45. Burettes for quantitative neutralization of acids and bases

Experiment 72. Prepare about half a liter of a 5% solution of caustic soda. Dilute 100 ccm. of hydrochloric acid with about five times its volume of water. Fill a burette with each of these solutions and measure into a beaker 20 ccm. of the caustic soda solution. Add a few drops of litmus and then run in hydrochloric acid from the second burette until the blue color of the litmus just changes to red. The solution must be stirred with a glass rod while the acid is being added, and the last few drops should be run in slowly and carefully. If a little too much acid is added and the liquid becomes suddenly red, it can be brought back to the neutral point by adding sodium hydroxide solution drop by drop. When the solution is neutral a single drop of acid should make it red and a single drop of alkali should make it blue.

89. Standard solutions. For convenience and rapidity of analysis it is usual to prepare solutions of acid and alkali such that 1 ccm. of one solution will just neutralize 1 ccm. of the other. Suppose we find in Experiment 72 that 20 ccm. of the alkali requires 14.3 ccm. of acid for neutralization. In order that 1 ccm. of each may be equivalent to 1 ccm. of the other, we must dilute the acid to a volume which is $\frac{20}{14.3}$ of its present volume.

Fig. 46. The meniscus at the surface of the liquid in a burette

Experiment 73. Calculate from the data of the last experiment how much water you should add to your acid to make 1 ccm. of it exactly equal to 1 ccm. of the alkali. Measure the volume of the acid in a measuring cylinder and dilute it to the required amount; then test experimentally the accuracy of your calculation and measurement. If the two solutions are not exactly equivalent, cubic centimeter for cubic centimeter, recalculate the amount of water necessary to make them exactly equal, add the water, and redetermine the point of neutralization. If the acid has been diluted too far, a few drops of the stronger acid used in making up the solution must be added to bring it up to the required strength.

What is called a *standard solution* of sodium hydroxide contains 40.06 gm. of pure sodium hydroxide in a liter of solution, and a solution of hydrochloric acid which is *equivalent,* that is, 1 ccm. of which will just neutralize 1 ccm. of the sodium hydroxide solution, contains 36.46 gm. of pure hydrochloric acid in a liter of solution. If instead of hydrochloric we use sulphuric acid, the equivalent solution contains 49.04 gm. of pure sulphuric acid, in a liter of solution; the equivalent solution of nitric acid contains 63.05 gm. in a liter of solution. These numbers give the relation by weight in which these acids neutralize caustic soda.

If we take any other base and find the number of grams per liter needed to make a solution which will neutralize our hydrochloric acid solution, cubic centimeter for cubic centimeter, we find that 1 ccm. of the standard sulphuric or the standard nitric acid will also neutralize 1 ccm. of this new alkaline solution. In the same way, by finding how many grams per liter of a new acid are needed to neutralize, cubic centimeter for cubic centimeter, one of our alkaline solutions, we know that it will neutralize any other standard solution of alkali in the same way.

90. Equivalent weights. If we test all the acids and bases we have, finding out how much of each acid is needed to neutralize 40.06 gm. of sodium hydroxide, and how much of each base is needed to neutralize 36.46 gm. of hydrochloric acid, we can state all our results in a table which will show what are called the *equivalent weights* of acids and bases.

TABLE OF EQUIVALENT WEIGHTS

Acids		*Bases*	
Hydrochloric	36.46	Sodium hydroxide	40.05
Sulphuric	49.04	Potassium hydroxide	56.16
Nitric	63.05		

91. Combining weights. Not only the acids and bases but any substance whatever has a definite combining weight, and *when substances combine to form a chemical compound they always do so in the proportions of their combining weights.* This is called the law of combining weights. These weights are really proportional weights, and a number has been chosen to represent the combining weight of one single substance, all the others being based on this unit. The combining weight of oxygen is arbitrarily chosen as 16.00. In our choice of weights of the acids and alkalies we used 40.06 gm. of caustic soda as a basis for the calculation of all the other numbers, and we chose this particular value because 40.06 gm. of caustic soda contain exactly 16.00 gm. of oxygen.

It would be easy to find the combining weight of mercury by an experiment like Experiment 26, for we used there a compound of mercury and oxygen, decomposing it by heat, driving off the oxygen, and leaving the mercury behind. If we had taken 10.00 gm. of the mercuric oxide, and weighed the mercury which remained after all the oxygen had been driven off, we would find 9.2592 gm. of mercury. The difference would be the weight of oxygen in 10.00 gm. of mercuric oxide,— 0.7408 gm.,— and from this data we could calculate the relation between the combining weights of mercury and oxygen:

$$x : 16.00 = 9.2592 : 0.7408$$

x is the combining weight of mercury. It is 200.0

The law of combining weights makes the quantitative part of chemical reactions very easy to study and remember; for when we know the combining weights of the elements we can find the combining weights of all the compounds by simple addition. No matter how complex the compound may be or how peculiar the reaction used in making it, each element retains its combining weight. This law tells us immediately just how much of one substance must be used to combine with a certain weight of another substance. If we use too much of one or the other, the result is simply that some of one of the original substances is left over when the reaction is finished.

In order to understand and describe the quantitative part of chemical reactions we must know the combining weights of the elements, and the following table contains them for the elements we have already studied something about:

COMBINING WEIGHTS OF ELEMENTS

Oxygen	16.00	Chlorine	35.45
Hydrogen	1.01	Sodium	23.05
Nitrogen	14.04	Calcium	40.1
Carbon	12.00	Mercury	200.0
Sulphur	32.06		

92. The oxygen unit. The choice of 16.00 for the combining weight of oxygen is a purely arbitrary one, and it makes no difference in the results of our chemical analyses whether this or any other number is taken as the basis of calculation. There are reasons why oxygen has been chosen as the standard. It forms compounds directly with a great many other elements, so that the relation between the combining weight of oxygen and the other elements can be easily determined. Another reason is that the combining weights are in most cases nearer to whole numbers when oxygen is taken as 16.00 than when any other number is taken as a basis.

93. The determination of combining weights. Combining weights can be determined by roundabout methods as well as by direct comparison with oxygen. For example, the combining weight of sulphur can be obtained from the compound mercuric sulphide. It is found by analysis that mercury and sulphur are present in this substance in the proportion of

$$1 : 0.1603.$$

From this data we can calculate the combining weight of sulphur by the formula

$$1 : 0.1603 = 200.0 : x,$$

and from this, x is equal to 32.06. The combining weight of sulphur can also be found by direct comparison with oxygen, and exactly the same number is found in this way.

94. The atomic hypothesis. The facts about the conservation of the elements which are stated in sections 33 and 34, and the fact that an element is always present in a compound in the proportion of its combining weight, are very remarkable indeed. They are real *facts*, but they have suggested what is called the *atomic hypothesis*.

Suppose that each element is composed of *atoms*. These are particles so small as to be beyond the reach of all direct investigation. Suppose also that the atoms of each element are all exactly alike in all of their properties, especially in all having the same weight. These atoms are supposed to unite to form compounds, and to separate when a compound is broken up into elements. Each atom of an element has the same definite weight, and it therefore follows that in all the combinations and decompositions into which the element enters, it always appears in a relation by weight which corresponds to the weight of one of its atoms. The general laws which we have stated as facts fit this hypothesis perfectly

JOHN DALTON (1766-1844)

Founder of present atomic theory as applied in chemistry. Expanded previous theory to include the idea of multiple proportions well, and it gives us a picture of what might be if such atoms existed. The assumption that the atoms are so small as to be beyond the reach of direct investigation excludes the possibility of testing the hypothesis any further.

95. Atomic weights. The combining weights are often called *atomic weights*, and the hypothesis of atoms brings with it much convenience in expressing chemical reactions and the relations by weight of the various elements which enter them. We indicate by the letter O one combining weight (or one atom) of oxygen; by the letter S one combining weight (or one atom) of sulphur; by the letters Hg one combining weight (or one atom) of mercury; and these weights are in the relation

$$16.00 : 32.06 : 200.0.$$

By these *symbols*, as they are called, we represent elements, and by combining them we can represent compounds, not only showing what elements they contain, but also showing the relation by weight in which the elements are present in the compound.

The expression HgO indicates the compound mercuric oxide. It expresses the fact that mercuric oxide consists of mercury and oxygen and nothing else, and the further fact that it contains these two elements in the proportion

200.0 parts of mercury to 16.00 parts of oxygen.

The expression HgS indicates the compound mercuric sulphide, and shows that this substance consists of the elements mercury and sulphur in the proportion by weight

200.0 parts of mercury to 32.06 parts of sulphur.

The actual facts are of course expressed by the law of combining weights, and the hypothesis of the existence of atoms helps us only to remember these facts, or perhaps by suggesting the discovery of new ones.

The table inside the front cover of the book gives the atomic weights of all the known elements, with their names and the symbols which we use for the sake of abbreviation in describing compounds and reactions. These symbols are usually the first letters of the names of the elements, but some are the first letters of the Greek and Latin names and not the English ones. Na is the symbol for sodium, and it is taken from the Latin name *natrium;* K, the symbol for potassium, is from the Latin name *kalium;* Fe, for iron, is from *ferrum;* W, the symbol for tungsten, is from the German name Wolfram. There are so many elements that it is of course impossible to use single letters for all of them, and so the most important, or the one first discovered, usually takes the single letters for its symbol, while the less common ones, or those of later date, have two letters in their symbols.

96. Chemical reactions. Not only can these symbols be used to represent elements and compounds, but they can also be used in the complete expression of a chemical reaction. The equation

$$Hg + O = HgO$$

indicates that mercury and oxygen combine to form mercuric oxide, and that 200.0 parts by weight of mercury combine with 16.00 parts by weight of oxygen, giving 216.0 parts of mercuric oxide.

Such a chemical formula must always contain the same symbols and the same number of each symbol on the two sides of the equality sign. This is merely another way of stating the law of the conservation of the elements, that is, of saying that no amount of an element can ever be created or destroyed. It is also an expression for the law of the conservation of mass, for each symbol represents a definite quantity of an element, and if we could have less of one symbol on one side of the equation than on the other, this would mean that mass had been created or destroyed.

97. Combination in more than one proportion. In what we have seen and learned so far we have taken no account of the possibility of more than one compound being formed from the same elements. We have, however, studied two compounds of carbon and oxygen,— carbon monoxide and carbon dioxide,— and the symbols for these two substances are CO and CO_2. In the first, carbon and oxygen are combined in the proportion

$$12.00 : 16.00.$$

This is carbon monoxide.

Carbon dioxide contains more oxygen in the same weight of gas, and the two elements must be present in it in some other proportion. Analysis shows that carbon dioxide contains 12.00 parts by weight of carbon to 32.00 parts of oxygen. Its formula is therefore CO_2.

98. Multiple proportions. It is quite possible for elements to combine in more than one proportion by weight, but experiment has shown that they combine in all cases either in the relation of their combining weights or in the relation of simple multiples of these weights. Two elements can combine in the proportions

$$1:1,\ 1:2,\ 1:3,\ 2:3.$$

Three elements can combine

1 : 1 : 1, 1 : 1 : 2, 1 : 2 : 3, etc.

The numbers which show how many times the combining weight enters the compound are seldom over 6 or 7, and so the number of possible chemical compounds is very large, without any very great complexity being added to the study of the relations between them. We can be sure beforehand that the relations by weight of elements in compounds will always be simple multiples of their combining weights.

The following table shows the combining weights of a number of compounds:

SYMBOLS AND COMBINING WEIGHTS

Water, H_2O	18.02	Sodium hydroxide, NaOH	40.06
Carbon monoxide, CO	28.00	Potassium hydroxide, KOH	56.16
Carbon dioxide, CO_2	44.00	Sodium chloride, NaCl	58.50
Hydrochloric acid, HCl	36.46	Silver nitrate, $AgNO_3$	169.98
Sulphuric acid, H_2SO_4	98.08	Calcium sulphate, $CaSO_4$	136.18
Nitric acid, HNO_3	63.05		

99. The determination of combining weights. These numbers have been experimentally determined by analyzing the compounds and finding the proportions by weight in which the elements are present. From this data and the combining weights themselves we can calculate the number of times the combining weight of each element is present in the compound. Analyses of this sort are sometimes very simple, as in the case of mercuric oxide, but in other cases the analysis is more difficult and a roundabout method may be necessary.

Experiment 74. Weigh out 10 gm. of black oxide of copper and transfer it carefully to a hard-glass tube arranged as in Figure 47. Generate hydrogen gas as in Experiment 42, dry it by passing it through a tube containing lumps of calcium chloride, and pass the current of dry hydrogen over the oxide of copper, heating it at the same time with a burner. At the far end of the hard-glass tube a weighed drying tube is attached, and the gases which leave the tube all pass through it. Whatever water is formed by the reaction will be collected and held by the calcium chloride in this weighed tube. In order that no water may condense in the hard-glass tube it must be kept warm along its whole length.

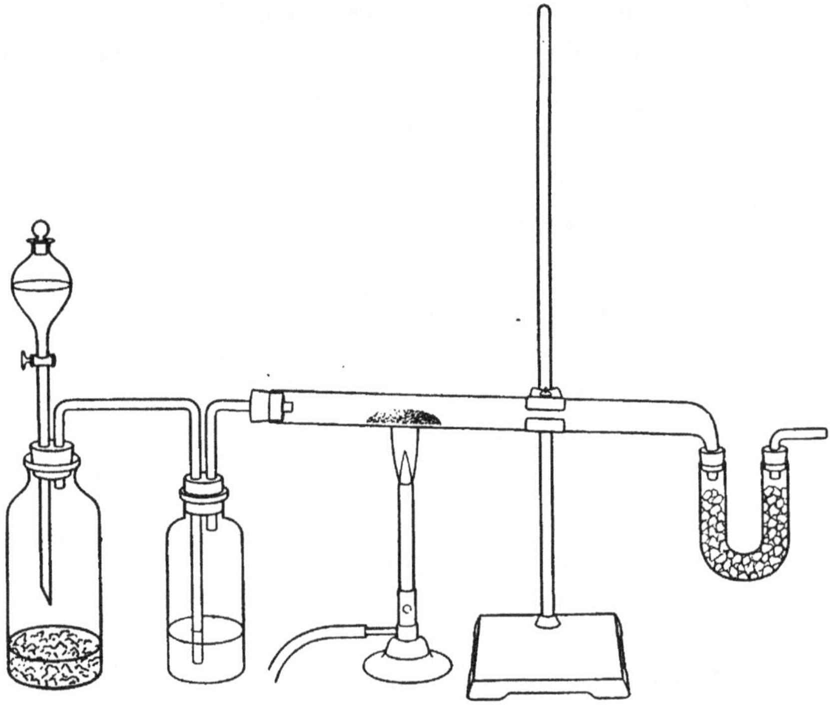

FIG. 47. Apparatus for the determination of the formula for water and the formula for copper oxide

The hydrogen combines with the oxygen of the copper oxide to form water, and the loss of weight of the copper oxide gives us the amount of oxygen in the water produced.

The difference between this weight of oxygen and the total weight of water produced gives the weight of hydrogen in this same weight of water. This gives all the data necessary for the calculation of the composition of water.

Experiment 74 *(continued)*. Allow the hydrogen gas to pass until the copper oxide is completely reduced to copper. Then calculate the composition of water as explained above.

Calculate also the composition of copper oxide.

The calculation will have the form :

Weight of copper oxide taken . . . .
Weight of copper left . . . .
Weight of oxygen . . . .
Weight of drying tube before experiment . ...
Weight of drying tube after experiment . . ..
Water produced . . . .
Weight of hydrogen (weight of water – weight of oxygen) ....
Water contains ....% oxygen and ....% hydrogen.

The combining weight of oxygen is 16.00, that of hydrogen is 1.01. The formula for water is therefore ....

The combining weight of oxygen is 16.00, that of copper is 63.6. The formula for copper oxide is therefore ....

100. The choice of combining weights. When we find more than one compound of two elements we must make a choice of the combining weight which we shall give to each. We have already learned about two compounds of carbon and oxygen, carbon monoxide and carbon dioxide. If we have fixed the combining weight of oxygen as 16.00, we must now choose which of the two compounds is to have *one* combining weight of carbon in its symbol. If we choose carbon monoxide, we find that it contains 12.00 parts of carbon to 16.00 of oxygen, and the combining weight of carbon is 12.00. If we choose carbon dioxide, we find that it contains

6.00 of carbon to 16.00 of oxygen, and the combining weight of carbon would be 6.00. In the first case the formula for carbon monoxide will be CO and that for carbon dioxide CO_2. In the second case the formula for carbon monoxide will be C_2O and that for carbon dioxide CO. A study of these two compounds alone will not give us any indication as to which of the two sets of symbols we should choose, but a study of the other compounds of carbon leads us to choose 12.00 as the combining weight of carbon because it gives the simplest set of symbols for all of them.

Since the idea of the combining weights was introduced into chemistry a hundred or more years ago, many changes in their values have been made, and some of them have been doubled to fit the facts better as more and more compounds became known. A glance at the table of combining weights in an old chemistry will show many such changes.

QUESTIONS

1. What is meant by a standard solution of an acid or an alkali?
2. If we assume for an acid or a base the number of grams which we will dissolve in a liter of solution, how can the equivalent weight of all other acids and bases be determined?
3. What substance is chosen as a basis for the combining weights?
4. Make as complete as possible a statement of the various uses to which the law of combining weights can be applied.
5. Assuming the combining weight of oxygen to be 16.00, how would you find the combining weight of mercury?
6. State the facts you have learned about combining weights in terms of the atomic hypothesis.
7. What does the formula HgO indicate?
8. How are the symbols which represent elements chosen?
9. If two elements combine in more than one way, how is this fact expressed in symbols?
10. State the law of multiple proportions.

11. How did you determine the proportion by weight in which hydrogen and oxygen are present in water?

12. How did you determine the formula for water?

13. How did you determine the formula for copper oxide?

14. If two elements combine in more than one proportion, what determines the choice of the combining weights?

PROBLEMS

1. What weight of mercury could be obtained from 100 gm. of mercuric oxide?

2. What weight of oxygen would be obtained?

3. What weight of sulphur must be used to make 100 gm. of mercuric sulphide?

4. What weight of water will be produced by reducing 100 gm. of black oxide of copper by hydrogen?

5. What weight of oxygen will be contained in this amount of water?

6. What weight of copper can be obtained from 100 gm. of the black oxide of copper?

CHAPTER XII

ELECTROLYSIS

101. The electrolysis of water. In Experiment 74 we *synthesized* water by causing hydrogen to combine with the oxygen which it took from copper oxide, and by the law of constant proportions and the law of combining weights we found the proportion by weight in which hydrogen and oxygen are present in the substance water. With the aid of an electric current it is quite easy to break water up into its elements and study its composition by *analysis.*

We know already that when hydrogen burns in oxygen heat is given out, and from this we may be quite sure that it will take work to break water up into hydrogen and oxygen. It is possible to break water up by heating it to a very high temperature indeed, but it is very much easier to do it by means of an electric current, which furnishes a supply of energy and can be used for the decomposition of many other substances beside water.

Experiment 75. Arrange an apparatus as in Figure 48, filling the tubes with dilute sulphuric acid and using two or three storage cells or four dry cells as a source of electrical energy. Allow the current to pass until the tubes are nearly filled with gas.

In what proportion by volume are the gases formed? Test the gas in each tube with a glowing splinter. What gases have been produced? What is the composition of water by volume?

102. Galvanic cells. This is the first occasion we have had to use any form of energy other than heat to bring about chemical changes, and the electrical current in this case is produced by a chemical reaction and at the expense of chemical energy in the cells. Many chemical reactions can be made to produce a current of electricity, and arrangements of this kind are called *galvanic cells.*

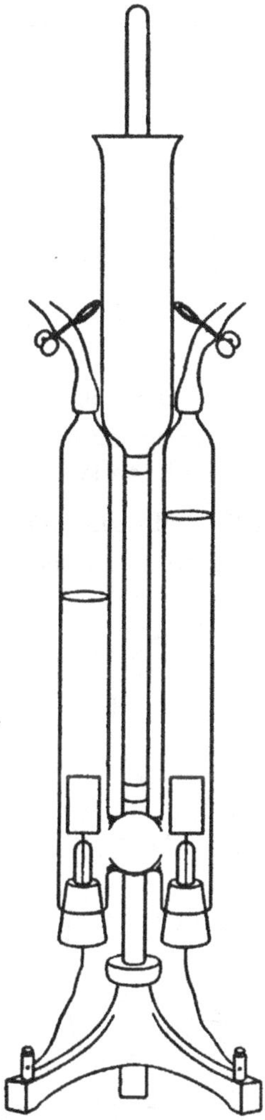

FIG. 48. Apparatus for the electrolysis of water

Experiment 76. Dip a zinc rod into a concentrated solution of copper sulphate. Try rods of magnesium, lead, and iron in the same way.

Dip a stick of zinc and a piece of copper into dilute sulphuric acid. What happens? Connect the two metals by bringing them into contact above the level of the acid.

Experiment 77. Make up a galvanic cell as in the figure. The outer jar contains a plate of zinc in dilute sulphuric acid. The porous cup contains a plate of copper in strong copper sulphate solution.

Arrange two of these cells so that the zinc of one cell is connected to the copper of the next. The two end wires, the one from the copper at one cell and the one from the zinc at the other, are to be connected to the apparatus of the next experiment.

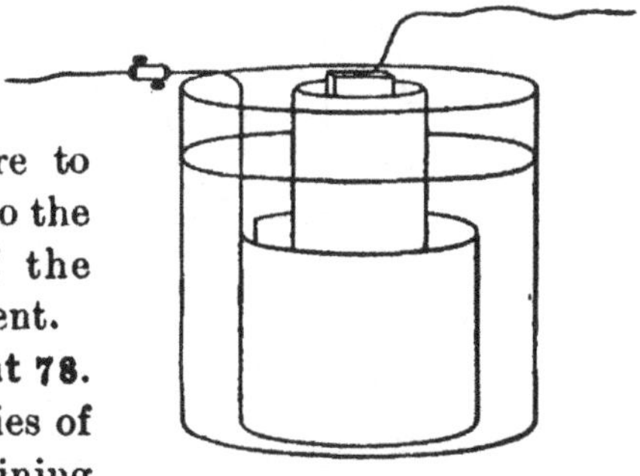

FIG. 49. Daniell cell

Experiment 78. Prepare a series of glasses containing the following solutions: (1) copper sulphate, (2) silver nitrate, (3) sulphuric acid (dilute), (4) potassium iodide.

Prepare copper *electrodes* — two for each glass. These are strips of copper about an inch wide and three inches long, with a wire soldered at one end. Connect the electrodes, a pair at a time, to the end terminals of your pair of galvanic cells. Let the current pass through each cell for several minutes.

Write out a complete statement of all the results obtained. The electric current has the power to cause a great many chemical changes. We found in Experiment 75 that it could decompose water into its elements, and we now see that it can set free other elements from their compounds.

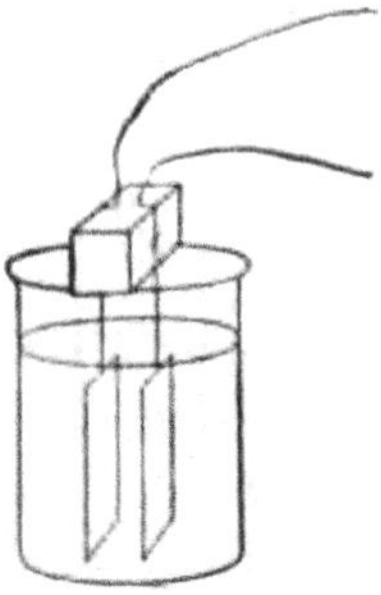

FIG. 50. Arrangement for electrolysis

Experiment 79. Insert a galvanometer, which is an instrument used to measure the strength of an electric current, in the circuit to one of the cells used in the previous experiment. How much is the needle deflected by the current?

Prepare a strong solution of sugar in distilled water. Place copper electrodes in this solution and attach to the two galvanic cells, with the galvanometer in the circuit. How much is the needle deflected under these conditions? See Figure 51 for a plan of the connections in this experiment.

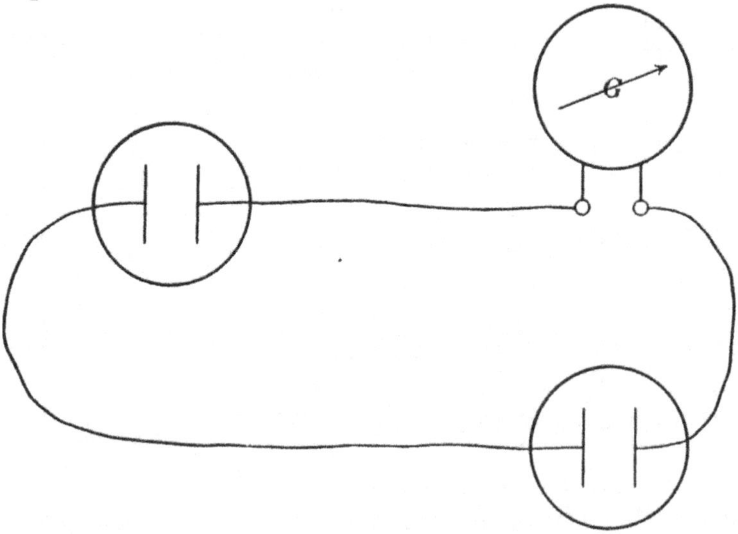

FIG. 51. A galvanometer in the circuit

103. The conductivity of solutions. Substances in solution differ very greatly in their power of conducting or carrying an electric current. Salts conduct well; sugar and most organic substances conduct only very slightly. The galvanometer in the above experiment gives an approximate measure of the conductivity of the solutions used.

104. Beside the qualitative results which we have found as the result of Experiment 78, there are some very simple quantitative relations which connect the amounts of substances which are dissolved or precipitated by an electric current with their combining weights.

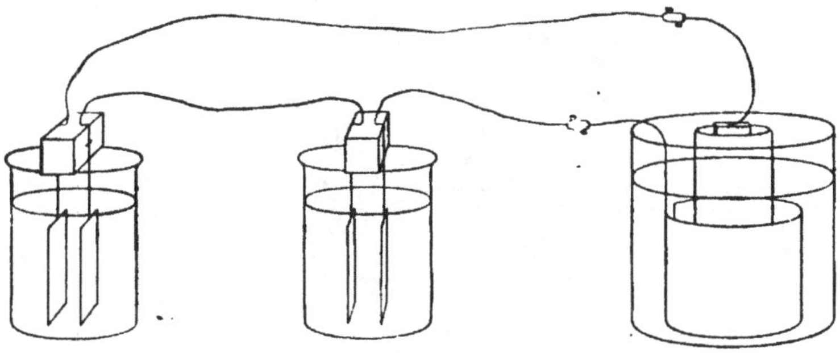

FIG. 52. Silver and copper voltameters

Experiment 80. Weigh carefully two copper and two silver electrodes with their binding screws. Place the copper electrodes in a strong solution of copper sulphate and the silver electrodes in a strong solution of silver nitrate (see Fig. 52). Let the current from your two galvanic cells pass through the two solutions for an hour, the connections being made as in the figure. Take out the electrodes and dry them by dipping them into alcohol and touching a match to them. Then weigh them again accurately.

What relation is there between the loss of weight of one electrode of each pair and the gain of weight of the other?

What relation is there between the amount of copper dissolved and deposited and the amount of silver dissolved and deposited?

If you see evidence of a general law, state it as clearly as you can.

105. Terms used in electrolysis. In order to describe what takes place when a current of electricity passes through

MICHAEL FARADAY (1791-1867)

tablished electrochemistry by the discovery of Faraday's Law. A pioneer and an experimental genius in many branches of physical science a solution, Faraday, in 1834, found it necessary to give names to a number of things which had not had names up to that time. He called a substance which allows a current of electricity to pass through it, with a resulting decomposition, an *electrolyte.* The solutions of salts and acids which we used are electrolytes. The process by which a current passes through an electrolyte he called *electrolysis.* He called the places where the current enters and leaves the electrolyte the *electrodes,*— the one where the current enters, that is, the one where a metal goes into solution, he called the *anode,* and the other, the place where the metal goes out of solution, the *kathode.* The substances which pass through the solution to the electrodes he called *ions,*— the one which moves through the solution toward the anode, the *anion,* the one which moves toward the kathode, the *kation.*

In section 85 we used these same names for the parts of salts in solution, and it has turned out that these same ions which carry the current during electrolysis are the ones which give its properties to a salt solution. Silver ion and nitrate ion determine the properties of a solution of silver nitrate, and silver ion moves to the kathode during electrolysis and is changed to metallic silver when it reaches it. At the same time nitrate ion moves toward the anode and takes silver from the anode, where it changes from metallic silver to silver ion. In the case of copper we could see by the blue color (Exp. 78, (3)) that copper ion is being formed at the anode.

106. The reaction in the Daniell cell. The cells which were used as a source of electricity in Experiments 77, 78, and 79 are of the type known as Daniell's cell. The chemical reaction which goes on while such a cell is furnishing a current is a very simple one. Zinc goes into solution and copper goes out. Exactly the same process takes place when a stick of zinc is dipped into a solution of copper sulphate, but in this latter case the copper is precipitated on the zinc, while in the Daniell cell we arranged for the zinc to go into solution at one place and for the copper to go out at another. We took pains to prevent any of the copper from reaching the zinc by means of the porous cup.

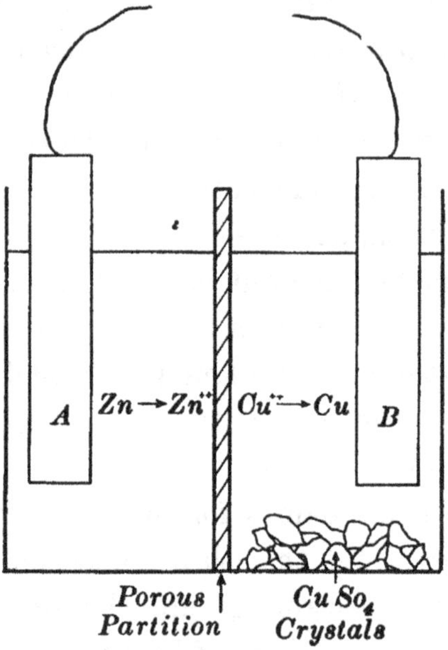

FIG. 53. Diagram of reaction in the Daniell cell

When the copper precipitated on the zinc, all of the chemical energy which was set free by the reaction was changed into heat. By separating the two reactions the chemical energy of the process can be changed into electrical energy, which can then be carried along wires and used whenever we like.

107. Secondary reactions. We found in earlier experiments that the ions of sodium hydroxide are

sodium ion,
hydroxide ion.

It so happens that neither of these ions can exist in water in the ordinary nonionic condition. Sodium reacts with water as soon as it is formed from its ion, forming sodium hydroxide, which remains in the solution, and hydrogen gas, which escapes. The hydroxide ion breaks up into water and oxygen gas if we attempt to free it from the ionic state.

Experiment 81. Electrolyze a solution of sodium hydroxide in an apparatus like that used in Experiment 75. What are the products of electrolysis in this case? What is the composition of sodium hydroxide?

A reaction like this, in which the ions which carry the current through the solution do not appear at the electrodes, but react there to form something else, is called a *secondary reaction.*

108. Summary. We may summarize the facts we have learned from a study of electrolysis as follows:

1. Acids, bases, and salts in solution conduct electricity and are decomposed when the current passes. Other substances like sugar, which form solutions equally well, do not conduct electricity in this way.

2. The passage of an electric current through an electrolyte is always accompanied by the transfer of a definite amount of two substances through the solution, one going in one direction and the other in the opposite direction.

3. Amounts of various substances which are dissolved or precipitated at the electrodes are in the proportions of their combining weights or simple fractions of these weights.

109. The hypothesis of electrolytic dissociation. Just as we added to the facts about the combining weights and the conservation of the elements the hypothesis that all substances consist of atoms, so we may add to the facts about ions in solution an hypothesis about what is called *electrolytic dissociation.* This is a sort of mental picture of what takes place when an acid, a base, or a salt is dissolved in water. It may be stated as follows:

When an acid, a base, or a salt is dissolved in water its particles each separate into two parts, or ions, each of which is accompanied by a definite quantity of electricity. One part of each is accompanied by positive electricity and the other by the same amount of negative electricity. The electric charges which are fastened to each of these particles are what make the ions so different from other substances, and it is only when an ion gives up its charge that it becomes like an ordinary substance.

The passage of an electric current through such a solution may then be imagined somewhat as follows:

An immense number of minute ions, each carrying a quantity of positive electricity, move through the solution toward the kathode. The same number of particles, each carrying the same quantity of negative electricity, move toward the anode. When the ions reach the electrodes they give up their charges, which pass away through the wires in the form of an electric current, while the ions themselves become atoms, and from this time on behave like any other chemical substance. If the ions are compound ones, like the sulphate ion or the hydroxide ion, a secondary reaction will take place at the electrode, the ions will be broken up, and other substances will appear.

Such a picture is beyond the reach of experiment, and we shall use Faraday's definition as given in section 105, which has no hypothetical bearing whatever. From our own experiments we can say that an ion is what moves through a solution toward either of the electrodes during the electrolysis of an electrolyte, and we can add to this what we learned from the study of solutions in earlier chapters.

QUESTIONS

1. Do you think it would require the expenditure of a good deal of work to separate water into hydrogen and oxygen? Why?

2. Describe the experiment in which water was electrolyzed into its elements.

3. How do salts and sugar in solution differ in their power of conducting an electric current?

4. Define the following terms: electrolyte, electrolysis, electrode, anode, kathode, ion, anion, kation.

5. What is the reaction which goes on in a Daniell cell while it is furnishing a current? Explain why this reaction can under some conditions give an electric current and under other conditions does not.

6. What substances are produced by the electrolysis of sodium hydroxide solution? What ions are present in this solution? Why do they not appear at the electrodes during electrolysis?

7. Summarize the qualitative and quantitative facts which you have learned from your study of electrolysis.

8. State the hypothesis of electrolytic dissociation.

PROBLEMS

1. Two electrolytic cells are connected in series. One contains copper sulphate solution and the other silver nitrate solution. The current is sent through both cells until 3.15 gm. of copper has been deposited. How many grams of silver have been deposited in the other cell?

2. A Daniell cell has been running until 32.5 gm. of zinc have been dissolved from the zinc electrode. How many grams of copper have been precipitated at the copper electrode?

CHAPTER XIII

GAS VOLUMES

110. Combination by volume. In the experiment on the electrolysis of water (Exp. 75) we found that 2 volumes of hydrogen were formed and 1 volume of oxygen. Since pressure and temperature have the same effect on all gases this ratio (2 : 1) will be the same under all circumstances. It is a fact that not only water but also all other compounds which are formed from gaseous elements behave in exactly the same way. Either equal volumes of gases combine to form them or result from their decomposition, or the volumes which combine have simple numerical relations to each other which are just the same as the relations of the combining weights.

If measurements are made at the same temperature and pressure, 1 volume of hydrogen combines with 1 volume of chlorine to form gaseous hydrochloric acid, and the volume of hydrochloric acid formed is exactly equal to the sum of the volumes of the two elements which united to form it. The law of combining weights is perfectly general and applies to gases as well as to solids and liquids, and since this is the case a very simple relation must hold between the combining weight of a gas and its volume.

111. Densities of gases. The following table gives the densities, in grams per cubic centimeter, of a number of substances, most of which are gases at ordinary temperatures, but including also water, or rather water *vapor*, which behaves like a gas at temperatures above 100° C., and at lower temperatures at pressures lower than one atmosphere:

DENSITIES OF GASES

Oxygen	0.00143	Nitrogen	0.00125
Hydrogen . . .	0.0000901	Carbon monoxide .	0.00125
Water	0.000804	Carbon dioxide . .	0.00196

Calculate how many cubic centimeters of each gas contains as many grams as equal the combining weight. This is done by dividing the combining weight by the density.

We were looking for a simple relation and it is now easy to see what it is. *The volume occupied by the combining weight of all gases is the same.*

11.2 liters of hydrogen (weight 1.01 gm.)

combines with

11.2 liters of chlorine (weight 35.45 gm.)

to form

22.4 liters of hydrochloric acid gas
(weight 36.46 gm.).

The simplest interpretation we could give to these facts in terms of the atomic hypothesis is to say that equal volumes of the gases contain the same number of atoms. When we try to apply this to the facts we find a contradiction: two volumes of hydrogen and one of oxygen combine to form two volumes of water vapor. If the number of atoms in the unit volume is N, then N atoms of oxygen and 2 N atoms of hydrogen combine to form 2 N atoms of water. There will then be only half an atom of oxygen in each atom of water, and half an atom is an idea which is contrary to our hypothesis.

112. The molecular hypothesis. To avoid this contradiction it seems best to assume that the smallest particles of gases are not single atoms but groups of atoms, which are called *molecules,* and this removes the contradiction in all known cases. Equal volumes of different gases are therefore assumed to contain the *same number of molecules,* and the molecules of the elements oxygen, hydrogen, nitrogen, chlorine, etc., have compositions O_2, H_2, N_2, Cl_2, etc. We can then express the relation in the case of water by saying that one molecule — two atoms — of oxygen combines with two molecules — four atoms — of hydrogen to form two molecules of water, each containing one atom of oxygen and two atoms of hydrogen.

This would be written in a chemical formula,

$$O_2 + 2\,H_2 = 2\,H_2O;$$

and the coefficients — 1 for the O_2, 2 for the H_2, and 2 for the H_2O — give us the volume relations directly, since what we have used as molecular quantities express quantities whose volumes (in the gaseous state) are equal. What is called the *molecular weight* of oxygen is 32.00, that of hydrogen is 2.02, and so on.

If we investigate all the elements, we find a few exceptions to the simple law that the molecules of any elementary gas consist of two atoms. Mercury vapor, for instance, has the density 0.00893, and since mercury has a combining weight of 200.0, the volume of 200.0 grams of mercury vapor is not 11.2 liters, but 22.4 liters.

We found that 11.2 liters is the volume of 1.01 gm. of hydrogen, and also of 16.00 gm. of oxygen. A molecule of hydrogen would correspond to double this weight, or 2.02 gm., and a molecule of oxygen to 32.00 gm. If we recalculate the table given in section 111 on the basis of *molecular* weights in place of combining weights, we find

O_2	22.4 liters
H_2	22.4 liters
Cl_2	22.4 liters, etc.

The number of grams of any gas which is equal to the molecular weight occupies a volume of 22.4 liters, no matter what its chemical nature may be, provided the gas is measured at 0°C. and under a pressure of 76 cm. of mercury. This number of grams is called a *gram molecule,* or sometimes simply a *mol.*

The formula for a molecule of mercury vapor must therefore be expressed by Hg and not by Hg_2.

113. Determination of molecular weight from density. It is therefore quite possible to calculate from the combining weight and the density of a gas the volume occupied by the combining weight, and it is also possible to reverse this process and calculate the molecular weight of a gas from its density in the gaseous state. The molecular weight in grams of any substance in this state occupies a volume of 22.4 liters. If we determine the density (as gas) of an unknown substance and find it to be, for instance, 0.00143 $\frac{\text{gm.}}{\text{ccm.}}$, we can calculate its molecular weight by multiplying this density by 22,400. This will give us the number of grams of the substance which would be contained in 22.4 liters. The calculation gives 32.00 gm., and we would recognize the substance as oxygen (O_2).

Just what the relation is between the molecular weight and the combining weight (or atomic weight) can only be decided by examining other compounds formed by the substance. An analysis by weight, together with a determination of the density in the state of gas, usually gives the best number to use.

114. The choice of combining weights. Water consists of 1.01 parts of hydrogen by weight to 8.00 parts of oxygen, as an accurate determination in Experiment 74 shows. The electrolysis of water (Exp. 75) shows that it contains two parts by volume of hydrogen to one part of oxygen. The simplest formula which fits these facts is H_2O, and in this the ratio of the atomic weights of hydrogen and oxygen is 1.01 : 16.00. The same formula follows from our table of densities of gases (sect. 111), and the fact that the molecular weight of any gas has a volume of 22.4 liters.

(Section 100 should be carefully considered in connection with the above statements.)

QUESTIONS

1. What is the composition of water by volume?

2. What is the volume occupied by the combining weight of a gas?

3. State the general law which tells how gases combine by volume.

4. State this same law in terms of the molecular hypothesis.

5. Why do we have to add the idea of molecules to our previous hypothesis of atoms in order to include the facts about gases?

6. How can the molecular weight of a substance in the state of gas be calculated from its density?

7. What further data may be necessary in order to find the combining weight (atomic weight) of the substance?

8. Express all the facts you have learned about gas volume by means of graphic statements, using the signs

[H_2] [O_2] [Cl_2] etc.,

for a *gram molecule* of hydrogen, oxygen, chlorine, etc.

9. How would you express the relation in the case of mercury vapor? Explain.

PROBLEM

1. From the table of section 111 calculate the molecular weight and the combining weight of each of the substances in the table.

CHAPTER XIV

THE HALOGEN GROUP

CHLORINE

115. The halogens. Certain groups of elements have properties which are very similar, or which show a regular gradation from member to member of the group, and the study of these elements is made easier by this fact. If we know the typical properties and reactions of the group, it is not hard to remember the differences between the various members.

One of the most interesting of these groups consists of the elements chlorine, bromine, and iodine, and these are called the "*halogens*," or salt formers, because they combine with all the metals to form salts. The members of this group have properties which are of course not the same, but which show simple gradations from element to element.

116. Chlorine. We have already used some of the compounds of chlorine, among them common salt and hydrochloric acid. Chlorine itself is a gas with properties more noticeable than those of any gas we have yet learned about. It has a greenish-yellow color and a very strong and unpleasant odor. In addition to this it combines with most of the elements very readily, even at ordinary temperatures, and it has the power of attacking many compounds and taking away one of the elements. It combines directly with hydrogen, and does it so readily in bright light that a sharp explosion is produced. What is usually called hydrochloric acid is a solution in water of the gaseous substance formed when hydrogen and chlorine unite.

Since chlorine combines so readily with many other substances it is never found free in nature. If it is produced in any way, it immediately combines with the nearest substance. Since it enters into combination so very readily, and usually with the evolution of a great deal of heat, we may be quite sure from our general rule that it will take a good deal of work to separate the element from its compounds and prepare it in the free state. It can be prepared from hydrochloric acid with the aid of a substance which combines very readily with hydrogen, and it is made, by what is called the Deacon process, by passing a mixture of hydrochloric acid gas and oxygen (or air) over red-hot pumice stone which has been saturated with copper sulphate. Water and chlorine are formed according to the reaction

$$2\,HCl + O = H_2O + 2\,Cl.$$

Or, if we wish to keep in mind the fact that the molecular weight of a gas is what enters into such reactions rather than the atomic or combining weight, we should write

$$4\,HCl + O_2 = 2\,H_2O + 4\,Cl.$$

Experiment 82. *Demonstration.* Air is blown through concentrated hydrochloric acid in the apparatus shown in Figure 54, and passes over pumice stone in the hard-glass tube. The pumice stone has been soaked in concentrated copper sulphate solution and then ignited to redness. Test the gas which escapes for chlorine. Allow the gas to bubble through water for a time, and test the water.

117. Catalysis. The copper does not appear in the chemical formula at all, and the reaction goes on without the presence of the copper salt, but very slowly indeed, so that only very minute quantities of chlorine are formed even after many hours. The presence alone of the copper salt hastens the reaction very greatly. It is called a *catalyser*, and we shall meet with other cases of this same accelerating action by substances which do not apparently enter the chemical formula. *Catalysis,* as this action is called, and *catalytic*

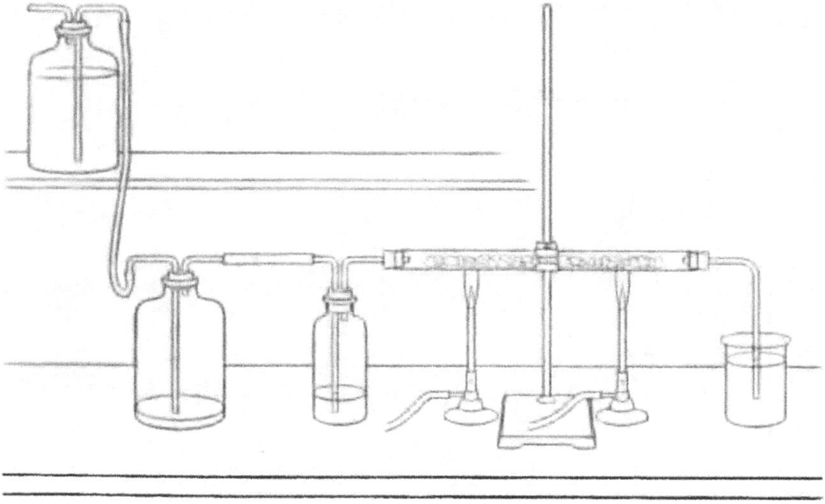

FIG. 54. Apparatus for the Deacon process

processes are of immense importance in technical chemistry, where it is a matter of economy to use reactions which take place as quickly and completely as possible.

118. The laboratory preparation of chlorine. This is not the usual way of making chlorine in the laboratory, where expense is not so important and where we try to avoid, as far as possible, complicated apparatus and high temperatures. Even here we depend on the power of oxygen to combine with the hydrogen of hydrochloric acid, but instead of using gaseous oxygen we make use of substances which contain a great deal of oxygen in combination and which give up a part of it readily. Such a substance is manganese dioxide, which is also used in the process of making oxygen from potassium chlorate.

Experiment 83. Arrange apparatus like that in Figure 55, using a water bath rather than an asbestos plate or wire gauze, as the materials used are liable to dry at the bottom of the flask and allow the heat to rise too high at that point. The temperature of boiling

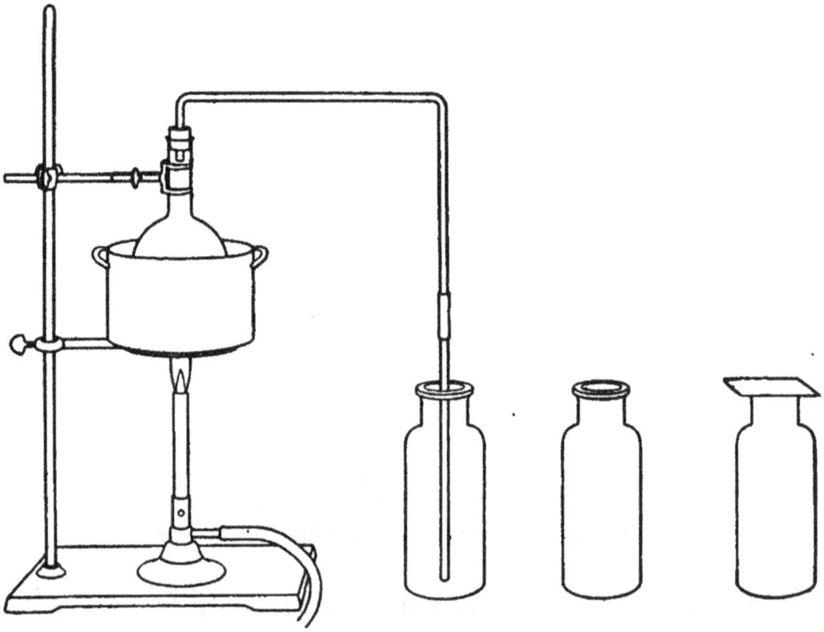

FIG. 55. Apparatus for the preparation of chlorine

water is quite high enough for the experiment, and it ensures that the bottom of the flask shall not be heated too hot.

Fill the flask about a third full of lumps of manganese dioxide in pieces as big as the end of the finger. Add enough strong hydrochloric acid to cover the pieces. Stopper the flask, arranging a delivery tube which reaches to the bottom of the collecting bottles, several of which should be provided. Heat the water bath and collect the chlorine in the receiving bottles, covering each bottle with a glass plate as soon as it is full.

NOTE. This experiment must be performed under a hood or in a window having a strong outward draft which will carry the fumes out of doors. Chlorine has a most irritating effect on the membranes of the nose and throat, and care must be taken not to inhale any of it. In case some is accidentally inhaled, a little alcohol, poured on the handkerchief and inhaled, is an antidote.

119. The properties of chlorine. The physical properties of chlorine are evident from this experiment. What are they? Is it heavier or lighter than air? The combining weight (atomic weight) of chlorine is 35.45, and its molecular formula is Cl_2. Calculate its density. What is its density compared to hydrogen?

Experiment 84. After enough chlorine for several experiments has been collected, — four or five bottles full, — pass the gas into water until a saturated solution is produced, and keep this solution in a stoppered bottle to be used in Experiment 95. Why could not chlorine very well be collected over water in the pneumatic trough?

Experiment 85. 1. Sprinkle some finely powdered metallic antimony into a bottle of chlorine.

2. Place a piece of sodium on an iron combustion spoon and lower it into a bottle of chlorine.

3. Dip a piece of filter paper into spirits of turpentine slightly warmed in an evaporating dish. Hold the prepared paper with forceps and lower into a bottle of chlorine.

4. Place a piece of dry, brightly colored calico in the gas. Try the same with a piece of wet calico.

State the results of all these experiments.

120. Chlorides. Chlorine combines directly with most metals to form *chlorides.* In naming them we place the name of the metal first and then the termination "chloride" (see sect. 84). We have just made antimony chloride and sodium chloride by causing chlorine gas to combine directly with the metals antimony and sodium. The reaction in the case of sodium is very simple:

$$Na + Cl = NaCl.$$

The reaction with antimony is a little more complicated:

$$Sb + 3\ Cl = SbCl_3.$$

Chlorine also combines very readily with the hydrogen of many organic compounds, and this is shown by the experiment with the turpentine, where the chlorine took the hydrogen away from its compound with carbon (turpentine

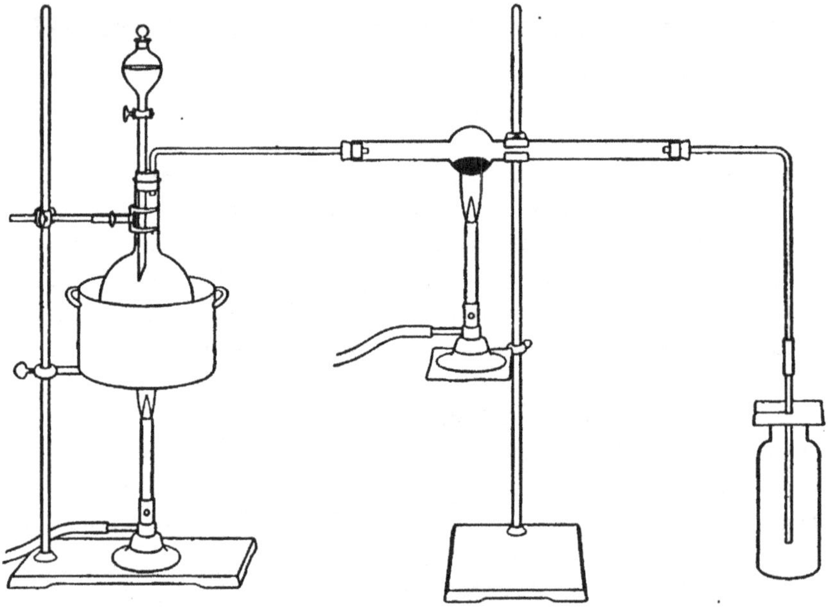

FIG. 56. Apparatus for making mercuric chloride from mercury and chlorine

is a hydrocarbon, a compound of carbon and hydrogen), leaving the latter behind in the free state as lampblack.

Most cheap dyes are compounds containing hydrogen, and chlorine breaks up these colored compounds and takes hydrogen away from them, especially when moisture is present, hydrochloric acid being formed during the process.

When a metal has the power of combining in several proportions with other elements, it can form more than one chloride.

Experiment 86. Pass dry chlorine gas over metallic mercury heated in a tube and arranged as in Figure 56. NOTE I The substance which is formed is extremely poisonous and should not be handled except by the teacher.

The chloride of mercury which is formed in this way is called mercur*ic* chloride.

121. *Ic* and *ous* compounds. When mercuric chloride is heated with mercury (without the presence of chlorine) another chloride of mercury is formed which is called mercur*ous* chloride. These two terminations are used to express the fact that one combining weight of mercury combines in one case with more chlorine than in the other. Analysis of mercuric chloride shows that its formula is $HgCl_2$, while mercurous chloride is $HgCl$. The ending *ic* is always used for the compound in which the metal combines with *more* chlorine, the ending *ous* where it combines with the *less* number of combining weights of chlorine. Where a metal combines with chlorine in only one proportion, as is the case with sodium and potassium, it makes no real difference how we name this chloride, but it is usual to retain the ending *ic* and call these compounds sodic chloride and potassic chloride. It does just as well to retain the ending *ium* of the metal, and so call the salts sodium chloride and potassium chloride.

122. Chlorine as a disinfectant. One of the principal uses of chlorine depends on the fact that it can combine so easily with the hydrogen of organic substances. It is used to destroy bad odors of organic origin and to stop the growth of living organisms. Used in this way as a disinfectant it destroys bacteria and other dangerous plant life by taking away hydrogen and so breaking up parts of the organisms which are necessary for their life processes. Chlorine gas is not often used for this purpose directly, for it has many disadvantages. Instead of the gas a compound is used which is made by passing chlorine over wet lime, and which is called bleaching powder or "chloride of lime." The chlorine in this substance is so loosely combined that it escapes slowly and performs the work of disinfection as well as the gas itself would, besides being easy and safe to handle and transport.

123. Bleaching action. Another important property, the ease with which chlorine combines with the hydrogen of organic substances in the presence of moisture, makes it useful in *bleaching*. For this purpose it is used in the form of bleaching powder, and its action has been illustrated in the case of the piece of calico.

When chlorine is passed into a solution of sodic hydroxide the reaction

$$2\,NaOH + Cl_2 = NaCl + NaClO + H_2O$$

takes place. The solution on the right-hand side of this equation is called "bleach," and it contains beside common salt the substance NaClO. This is a type of a new kind of compound, one containing a metal, chlorine, and oxygen, and it is called sodium *hypochlorite*. It is a salt of an acid which is called hypochlorous acid; its formula is HClO.

124. Oxygen acids of chlorine. I. Chlorine forms a number of acids which contain oxygen. They are

ACID	NAME	SALT	NAME	ION	NAME
$HClO$	Hypochlorous acid	$NaClO$	Sodium hypochlorite	ClO^-	Hypochlorite
$HClO_2$	Chlorous acid	$NaClO_2$	Sodium chlorite	ClO_2^-	Chlorite
$HClO_3$	Chloric acid	$NaClO_3$	Sodium chlorate	ClO_3^-	Chlorate
$HClO_4$	Perchloric acid	$NaClO_4$	Sodium perchlorate	ClO_4^-	Perchlorate

This table gives a key to the naming of substances which contain elements in more than one proportion by weight, and which contain oxygen as a part of the anion. If there are only two such compounds, they are called the *ate* and the *ite*, the former being the one with the larger proportion of oxygen. If there is also a compound containing more oxygen than the *ate*, it is given the prefix *per;* and if there exists another compound containing less oxygen than the *ite*, it is given the prefix *hypo*.

125. Symbols for ions. When ions are to be distinguished from other substances in a chemical formula we do this by adding the sign ($^-$) to negative ions, or anions, and by adding ($^+$) to positive ions, or kations. Cl means elementary chlorine, but Cl^- means chlorine ion; Na is metallic sodium, while Na^+ is sodium ion. Some ions unite with more than one combining weight of the simpler ions; sulphuric acid, for example, contains twice the combining weight of ionic hydrogen and its formula is H_2SO_4. Sulphate ion is therefore called a bivalent ion, and others are trivalent; so we attach plus signs or minus signs enough to show how many univalent ions, like hydrogen ion or chlorine ion, combine with one of the ions with higher valence. Sulphate ion is written SO_4^{--}. Other cases are

Calcium ion	Ca^{++}
Mercuric ion	Hg^{++}
Ferrous ion	Fe^{++}

These two signs follow a law of conservation in chemical reactions. Whenever an element (or a group of elements) is changed into an ion it takes up a definite quantity of electricity, and this may be transferred from one substance to another, so that the second element or group may become an ion at the same time that the original ion becomes an ordinary substance. When we place a stick of zinc in copper sulphate solution, as in Experiment 76, the chemical change which takes place is

$$Cu^{++}\ SO_4^{--} + Zn = Zn^{++}\ SO_4^{--} + Cu.$$

The sulphate ion SO_4^{--} does not change at all during this reaction, so we may leave it out, and the reaction is then simply

$$Cu^{++} + Zn = Zn^{++} + Cu.$$

This means that any solution which contains copper ion, treated with metallic zinc, will give metallic copper and zinc ion. The anion has no influence on the reaction.

Experiment 87. Test this statement by placing a zinc rod in each of the following solutions: cupric nitrate, $Cu(NO_3)_2$; cupric chloride, $CuCl_2$; cupric bromide, $CuBr_2$.

The reaction which we used as a test for an acid, that with metallic magnesium, may be written in the same way,

$$2\,H^{+} + Mg = Mg^{++} + H_2,$$

indicating that in every case where metallic magnesium meets with hydrogen ion it is changed into bivalent magnesium ion, Mg^{++}, twice the combining weight of hydrogen ion, H^{+}, giving up its electric charge and becoming gaseous or nonionic hydrogen.

Equal quantities of positive and negative electricity neutralize each other, and when an anion and a kation unite to form a neutral substance the $(^{-})$ which belongs to one and the $(^{+})$ which is always with the other both disappear.

$$Ag^{+} + Cl^{-} = AgCl$$

represents the precipitation of insoluble silver chloride from a solution where silver ion and chlorine ion meet. In the same way an equal number of plus and minus signs may originate. The equation

$$CaCl_2 = Ca^{++} + 2 \cdot Cl^-$$

represents the formation of calcium ion and chlorine ion from nonionic (solid) calcium chloride when it is dissolved in water. It must be remembered in writing formulæ that the total sum of plus signs must always be the same as the total sum of minus signs on both sides of the equation.

126. Oxygen acids of chlorine. II. The potassium salts of the acids of section 124 are analogous to the sodium salts, and chemically are almost exactly like them. In some cases they are less soluble, however, and they can therefore be more easily prepared.

Experiment 88. Pass chlorine gas into sodic hydroxide solution. Let the solution stand unstoppered until the smell of chlorine has disappeared. What reaction has taken place? Add sulphuric acid to the solution. What gas is set free?

Experiment 89. Make a concentrated solution of potassic hydroxide, 1 part KOH to 2 of water. Saturate with chlorine gas. How can you tell by watching the bubbles as they rise when the solution is *saturated?* Allow the solution to stand for some time undisturbed.

127. Potassium chlorate. The crystals which separate are potassium chlorate ($KClO_3$), the potassium salt of chloric acid. It is formed in two steps:

$$(1) \quad 2\,KOH + Cl_2 = KClO + KCl + H_2O.$$

The substances first formed by the action are potassium chloride and potassium hypochlorite, just as in the case of sodium (sect. 123). The hypochlorite breaks up almost immediately into potassium *chlorate* and potassium chloride:

$$(2) \quad 3\,KClO = KClO_3 + 2\,KCl.$$

The potassium chloride which forms is very much more soluble than potassium chlorate, so it remains in the solution while the latter separates out in the solid form.

If the intermediate step is omitted (it really does take place), the reaction may be written

$$6\,KOH + 3\,Cl_2 = 5\,KCl + KClO_3 + H_2O.$$

128. The effect of temperature on solubility. Another property of potassium chlorate enables us to separate it still more perfectly from the substances which are present with it in the solution. It is very much less soluble in cold water

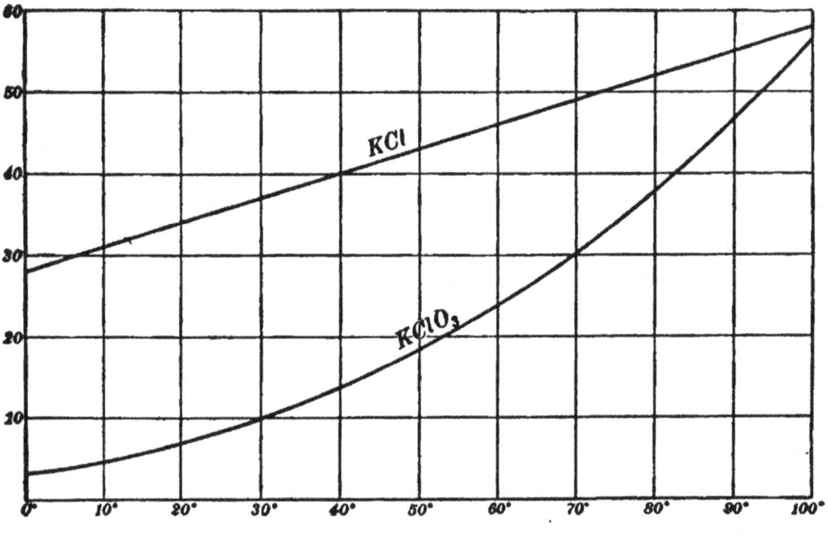

FIG. 57. Solubility curves of potassium chloride and potassium chlorate

than in hot, and when a solution containing much of the salt is cooled, crystals appear.

Experiment 90. Separate the potassic chlorate crystals from the solution by filtration. Remove as much of the adhering liquid as possible with dry filter paper and preserve the dry crystals.

A salt prepared in this way by crystallization and filtration is not quite pure. A little of the "mother liquor," as the solution out of which crystals form is called, always adheres to them, and this cannot be completely removed by filter paper. Salts are purified by what is called *recrystallization*; that is, by dissolving in pure water and allowing them to crystallize out again. A good deal of the salt is of course lost by this process, but this loss is necessary if we wish to get the substance in a pure state.

Experiment 91. Dissolve the potassium chlorate made in the last experiments in hot water, being careful to use just enough water to dissolve the salt. Cool the solution. Filter off the salt and dry as before. Preserve the purified salt in a labeled bottle or tube.

A solution of potassium chlorate contains the ions K^+ and ClO_3^- and must therefore, according to our general law, show the properties of these two ions. Although chlorine is present in the solution of the salt, it is not present as chloride ion, and the solution should not show any reaction for the chloride ion. The salt was, however, prepared from a solution which did contain chloride ion, and only when the purification has been very complete will the properties of this impurity disappear. The purity of your potassium chlorate can be shown by testing it with silver ion. If no precipitate of silver chloride appears, it is free from adhering potassium chloride.

QUESTIONS

1. What are the halogens? What does the word mean?
2. What are the properties of chlorine? How does it occur in nature? Why does it not occur free?
3. How may chlorine be prepared from hydrochloric acid?
4. Write the reaction for the Deacon process.

5. How is chlorine usually prepared in the laboratory?

6. What name is given to compounds which contain chlorine and a metal?

7. If chlorine combines with a metal to form more than one compound, how are the two compounds named?

8. Why is chlorine a good disinfectant?

9. What substance is used in place of pure chlorine as a disinfectant? Why can it be used in place of pure chlorine?

10. How does chlorine bleach organic coloring substances?

11. What oxygen acids does chlorine form?

12. How are the salts of these acids named?

13. Write the formulæ for the potassium salts of all these acids.

14. How is sodium hypochlorite made?

15. How is potassium chlorate made?

16. How are ions expressed in symbols to distinguish them from other substances?

17. Write the Daniell-cell reaction in terms of ions.

18. Why is potassic chlorate easier to prepare than sodic chlorate?

19. What ions are present in a solution of potassic chlorate?

20. Does a solution of potassic chlorate show any test for chlorine ion? Explain.

21. Explain how a salt is purified by recrystallization.

22. Have you any data which tells you anything about the solubility of silver chlorate? Explain.

PROBLEMS

1. What weight of chlorine will be produced for each 100 gm. of hydrochloric acid used up in the Deacon process?

2. In Experiment 85, part 2, what weight of sodium chloride is produced from each gram of sodium used?

3. What weight of antimony (in part 1 of the same experiment) must be used to produce 10 gm. of antimony trichloride ($SbCl_3$)?

4. What weight of chlorine must be used to produce 100 gm. of sodium hypochlorite? What volume (at 0° C. and 76 cm.) would this weight of chlorine occupy?

5. What weight of chlorine must be used to form 100 gm. of mercur*ic* chloride? to form 100 gm. of mercur*ous* chloride?

6. How many cubic centimeters of a normal hydrochloric acid solution will be needed to precipitate 10 gm. of silver ion [Ag^+] as silver chloride?

7. What weight of potassium chlorate will yield 10 liters of oxygen? The reaction is

$$2\,KClO_3 = 2\,KCl + 3\,O_2.$$

CHAPTER XV

BROMINE AND IODINE (FLUORINE)

129. Bromine and iodine. These two elements are so nearly like chlorine in all of their chemical properties that we may write most of their reactions by simply inserting Br and I in place of Cl in the reactions already studied.

130. Bromine. Although the properties and reactions of these three elements are very similar, we will find progressive differences. Bromine is a liquid, the only liquid element beside mercury. It is deep red in color, heavy and oily-looking, and of most unpleasant odor, attacking the membranes of the nose and throat viciously if even a very small amount is inhaled.

Experiment 92. Examine the physical properties of bromine carefully. Is it heavy or light? Examine the surface of the liquid where it comes in contact with the walls of the bottle. Does it curve up like water or down like mercury?

Bromine boils at 60° C., but sends out this brown vapor even at ordinary temperatures, just as water sends out its vapor, except that in this case the vapor is quite visible and produces a very unpleasant effect if breathed, while vapor of water is invisible and odorless. The combining weight of bromine is 80.00 and the formula for bromine vapor is Br_2. (What is its density compared with hydrogen? Compared with air?) Liquid bromine has a density of about 3.0 compared with water.

Bromine is nearly as active chemically as chlorine, combining directly with many of the elements just as chlorine does. It is soluble in water, but only in small amount.

131. Bromides. The salts which are formed when bromine acts on metals are called bromides, and solutions of these salts contain the ion of the metal and the bromide ion. The reactions of the bromide ion are very similar to those of the chloride ion.

Experiment 93. Try the reaction between silver nitrate solution and a solution of a soluble bromide. Explain.

Silver bromide is, like silver chloride, a substance which is only very slightly soluble in water. It differs slightly from silver chloride in color, being pale yellow, while the chloride is pure white.

Experiment 94. Prepare some bromine water by shaking up a little bromine in a pint bottle full of water. Place some in a test tube and add a little powdered metallic magnesium. From what you have seen, what can you say about the color of the bromide ion?

The reaction just studied is a very simple one indeed. It is

$$Br_2 + Mg = MgBr_2,$$

and this dissolves in the water with the result

$$MgBr_2 = Mg^{++} + 2\,Br^{--}.$$

Experiment 95. Pass chlorine gas into (or add chlorine water to) the solution of magnesium bromide formed in the previous experiment. The equation is

$$Mg^{++} + 2\,Br^{-} + Cl_2 = Mg^{++} + Br_2 + 2\,Cl^{-},$$

and since the Mg^{++} does not change during the reaction we could just as well write

$$2\,Br^{-} + Cl_2 = 2\,Cl^{-} + Br_2.$$

132. The preparation of bromine. This last experiment illustrates one of the methods of preparing bromine on a large scale. The mother liquor of many salt springs, left behind after the less soluble salts have crystallized out, contains magnesium bromide, which is a very soluble substance. Chlorine gas is passed into this mother liquor and the bromine is set free, magnesium chloride remaining in solution. The bromine is separated by distillation and condenses to a liquid in the cooled receiver.

Experiment 96. Pour a little carbon bisulphide into a test tube. Add a little potassium bromide solution and then add chlorine water, a drop at a time, shaking after each drop. Explain.

Bromine is much more soluble in carbon bisulphide or ether than in water, and the bromine which is set free distributes itself between the water and the carbon bisulphide in the ratio of its solubility, most of it going into the carbon bisulphide and only a little into the water.

133. Hydrobromic acid. Bromine combines with hydrogen, just as chlorine does, to form a compound of formula HBr, called *hydrobromic acid.* The reaction between the two elements is not so sudden, and there is no explosion when the vapors are mixed and exposed to bright light. The solution of gaseous HBr in water is what is usually called hydrobromic acid. If we try to prepare this acid by a reaction similar to the one used in the preparation of HCl, we find the difference in the properties of the two substances to be quite plain.

Experiment 97. Place some crystals of potassic bromide in a test tube and pour concentrated sulphuric acid over them. What is formed? Test the properties of the vapor by holding a wet glass rod in the vapor in the tube and then touching it to litmus paper.

We could produce hydrochloric acid by a similar reaction without any chlorine being formed, but hydrobromic acid is partially broken up with formation of free bromine. Pure hydrobromic acid is made in other ways.

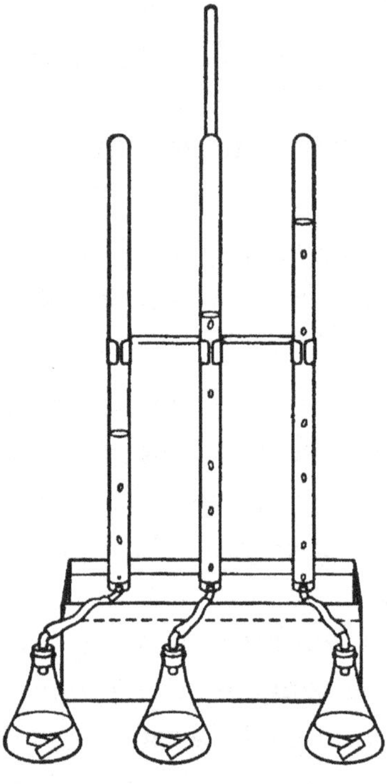

FIG. 58. Apparatus for showing the relative strength of acids

134. The strength of acids. Although hydrochloric acid is more *stable* than hydrobromic, the two are about equally *strong.* This means that a solution containing in one case 36.46 gm. of HCl per liter, and a solution containing 80.97 gm. of HBr per liter (each containing 1.01 gm. of hydrogen per liter), each contains about the same amount of *hydrogen ion.* This may be shown by allowing these two acids to act on zinc or magnesium and observing the rate at which hydrogen gas is evolved.

Experiment 98. Equivalent acids are prepared by titrating them with equivalent normal solution of sodic hydroxide and diluting or adding strong acid until the two acid solutions are exactly normal and exactly equal. Apparatus like that in the figure is arranged, a stick of zinc of the same size being placed in each flask. The acids are then added, an equal volume to each flask, and the evolution of gas allowed to go on for a time before collecting in the measuring tubes. The addition of a few drops of copper sulphate solution to each flask will make the evolution of hydrogen take place very regularly.

When we say that two acids are of the same strength we mean this very definite thing: that solutions of the two acids of the same equivalent concentration contain equal concentrations of hydrogen *ion.* When an acid is dissolved in water it does not change entirely into its two ions. A part remains un-ionized. The greater the fraction of the acid which has changed into ions the greater the concentration of hydrogen ion and the *stronger* the acid. The strength of an acid cannot be measured by neutralization or analysis. These processes show only the total amount of acid hydrogen and give no idea of the amount of hydrogen ion. When we neutralize an acid with a base, whatever hydrogen ion is present combines with the hydroxide ion of the base to form water. As fast as hydrogen ion is removed in this way more is formed from the acid, and the process goes on until all the hydrogen of the acid has combined with hydroxide ion, even though the concentration of the ions at any time was very small.

The velocity with which hydrogen is evolved from an acid by a metal is, however, a rough measure of the concentration of the hydrogen ion existing in the solution.

135. Potassium hypobromite. When bromine is added to a solution of potassium hydroxide, potassium hypobromite is formed, just as in the case of chlorine, and this substance is much like the corresponding chlorine compound. If concentrated potassium hydroxide solution is used and bromine added, potassium bromate is formed by a reaction quite analogous to the formation of the corresponding chlorate, and the salt separates out, its solubility being small. Here we have another similarity between compounds of chlorine and bromine.

136. Iodine. The third member of the family of halogens is a solid substance called *iodine.* Chlorine is a gas, bromine a liquid, and now we have a solid. So far they would seem to be as different as possible in their physical properties, but the state of a substance is something which is in very large measure dependent on the temperature, and if we compare these three elements at a temperature above 300° C., we shall find them very much alike indeed. Under these conditions chlorine is a greenish-yellow gas, bromine a reddish-brown gas, and iodine a purple gas. A still closer relationship appears if we examine their combining weights and densities, for bromine is very nearly the mean between the other two. Other properties show similar relations, as will be seen from the table.

	Cl	Br	I
Combining weight . .	35.45	79.96	126.85
Boiling point	– 33.6	63	184
Melting point	–102	– 7	114
Density (as gas) . . .	0.00317	0.00714	0.0113

Many of the chemical properties show gradations similar to these.

Iodine can be easily purified by *sublimation* (see sect. 14). It is heated and changed into a gas, and is then condensed on a cold surface, leaving behind it all nonvolatile impurities.

Experiment 99. Heat some iodine in a loosely stoppered test tube. Does it melt? What are the properties of iodine vapor, — color, approximate density, odor (carefully), action on metals, action on organic dyes, solubility in water?

Experiment 100. Test the reaction of a soluble iodide with silver nitrate solution. Compare with the reactions for chloride ion and bromide ion.

Experiment 101. Add chlorine water to a solution of a soluble iodide. Explain. Add bromine water to some of the same solution. Try also with the addition of a little carbon bisulphide.

137. Hydriodic acid. Hydrogen and iodine combine to form gaseous hydriodic acid, but the union between the two elements is feebler than in the case of hydrogen and bromine, and a mixture of the two elements only partially reacts even at high temperatures. Part of the hydrogen and iodine remains in the free state.

If we try to make the gas by the same method as was used for hydrochloric acid, only traces of acid are formed. It decomposes immediately into iodine, which appears in the free state, and hydrogen, which combines with the sulphuric acid.

Experiment 102. Place a few crystals of potassic iodide in a test tube and add concentrated sulphuric acid. Test the gas evolved for acid properties, as in Experiment 97. Can you detect any acid? What has been formed?

138. Hypoiodate and iodate. Iodine acts on potassic hydroxide solution just as the other halogens do, forming potassic hypoiodate, and this breaks up into potassic iodate and potassic iodide in a way analogous to the reactions of the corresponding chlorine and bromine compounds. In fact, the transformation into iodate takes place so rapidly that the hypoiodate cannot be kept for more than the shortest time.

139. The tests for free iodine. Iodine has one property which distinguishes it completely from the other halogens.

Experiment 103. Prepare a thin solution or paste of starch (see Appendix). Dissolve some iodine in water and add a few drops of the solution to the starch paste.

Starch is an exceedingly sensitive reagent for free iodine, and of course free iodine is an equally good reagent for starch. Very small amounts of the two substances show the blue color clearly.

Experiment 104. Add solution of potassic iodide to starch paste. Is starch a reagent for iodide ion? Add a drop of chlorine water to the mixture. Explain. Add a drop of bromine water in the same way.

Experiment 105. Fill a test tube with dilute starch paste. Add enough free iodine to turn it a strong blue. Heat the mixture. Cool the lower half of the tube, leaving the upper part warm.

The compound of iodine and starch is not very stable. It breaks up at higher temperatures, but reforms again when the temperature is lowered.

140. We are now ready to add some chemical properties to the physical ones given in the table of section 136.

The simple acids of the same type as HCl are about equally *strong*, but they show gradation in their stability. HCl is stable, HBr less stable, HI quite unstable.

The *hypo*-salts are similar in composition, but show a gradation in the velocity with which they go over into the *ates*.

Free chlorine drives either of the other halogens out of the ionic state, becoming chloride ion, according to the reaction

$$Cl_2 + 2\,Br^- \text{ (or } 2\,I^-) = 2\,Cl^- + Br_2 \text{ (or } I_2).$$

Free bromine can replace iodide ion in the same way.

141. Fluorine. This is a gas which belongs chemically to the family of the halogens, but which differs from them in so many ways that its compounds cannot be included in the typical classes we have found for them. It is chemically the most active of all the elements, and combines with everything so readily that it is only in comparatively recent times that the element has been prepared in the free state. It forms an acid, *hydrofluoric* acid, HF, which is a gas, easily soluble in water. This is used for etching glass, which it dissolves rapidly. The fluorides are almost all difficultly soluble substances, and not very much is known about fluoride ion.

QUESTIONS

1. What are the physical properties of bromine?

2. How would you test for bromide ion?

3. How is bromine prepared on a large scale?

4. If you had chloride ion and bromide ion in the same solution, how would you show the presence of each?

5. What is meant by the strength of acids? How can the comparative strength of two acids be found?

6. Why can you not determine the strength of an acid by titrating it with an alkali?

7. What are the physical properties of iodine? of iodine vapor?

8. How would you test for the presence of iodide ion?

9. If you had both iodide ion and bromide ion in the same solution, how would you prove the presence of each?

10. How would you test for the presence of free iodine?

11. Give a general statement of the properties of the three halogens, — chlorine, bromine, and iodine, — showing their relations and differences.

PROBLEMS

1. What weight of chlorine must be used to set free 100 gm. of bromine in Experiment 95?

2. What volume of chlorine will this be?

3. What weight of iodine will be set free in Experiment 101 by this same quantity of chlorine?

CHAPTER XVI

SULPHUR

142. Properties. Sulphur is an element with peculiar physical properties, and one which is of much importance in chemistry.

Experiment 106. Examine the physical properties of a piece of sulphur. Has it odor? taste? Is it soluble in water? in carbon bisulphide? Is it combustible?

Experiment 107. Heat some sulphur in an evaporating dish until it melts. Determine, by means of a thermometer, the point at which it solidifies again. Allow the sulphur to cool again and examine its properties. Heat to a point well above the melting point and cool again. Heat some sulphur in a hard-glass tube until it boils. What are the properties of the vapor? Pour the boiling molten sulphur into cold water. Examine its properties.

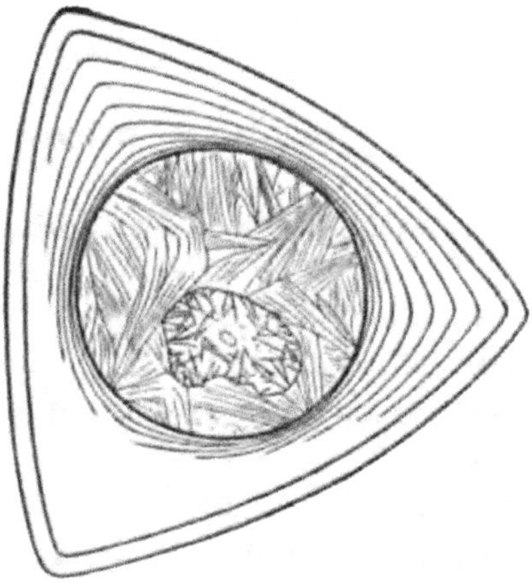

FIG. 59. Crystals formed by melting sulphur in a crucible

143. Allotropic forms of sulphur. The substance formed in the last part of Experiment 107 is another form of the element sulphur. It has no crystalline form and is called *amorphous* sulphur.

Just as the element carbon has three different forms,— amorphous (charcoal, lampblack, etc.), graphite, and diamond,

— so sulphur can also exist in several different allotropic forms, and amorphous sulphur, the substance we have just examined, is one of them. The element has also five or six different crystalline forms.

We can prove that these different forms all consist of elementary sulphur by examining the substances resulting from a chemical change of each of them, using the same weight for every form. Sulphur burns in air to sulphur dioxide gas, and if we burn equal weights of the different forms, the weight of sulphur dioxide gas formed in each case will be exactly the same.

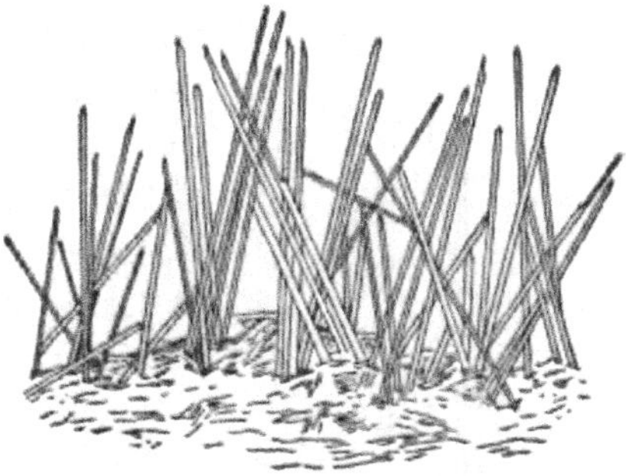

FIG. 60. Sulphur crystals

One of several crystalline forms which sulphur may show

There is, however, one very important difference in the different cases: different amounts of heat will be given out during the combustion. When the same weight of various allotropic forms of the same element enters a chemical reaction, the quantity of energy given out during the reaction is different for the different forms (see also sect. 65).

144. Natural occurrence of sulphur. The element sulphur occurs in large quantities in nature, usually in volcanic regions and especially in Sicily, Japan, and Mexico. It also occurs in combination in the *sulphides* of many metals. Lead sulphide is the mineral *galena*, ferric sulphide is a very common ore of iron, and sulphides of mercury and zinc are not uncommon minerals. The element is prepared by distillation from crude sulphur, which contains many impurities. Its boiling point is not very high (440° C.), and it can in this way be freed from all impurities which are not volatile at this temperature.

145. Sulphur dioxide. When sulphur burns in air or in oxygen, sulphur dioxide is formed according to the equation

$$S + O_2 = SO_2.$$

Experiment 108. Arrange an apparatus like that in Figure 61, using a flask with a capacity of two liters or more. Pass through the rubber stopper a manometer tube containing mercury, to show the variations of pressure within the flask, and pass also through the stopper a tightly fitting wire carrying a combustion spoon and reaching nearly to the bottom of the flask. Fill the spoon with sulphur, ignite it, lower into the flask, and stopper tightly. Describe results and explain them as fully as you can.

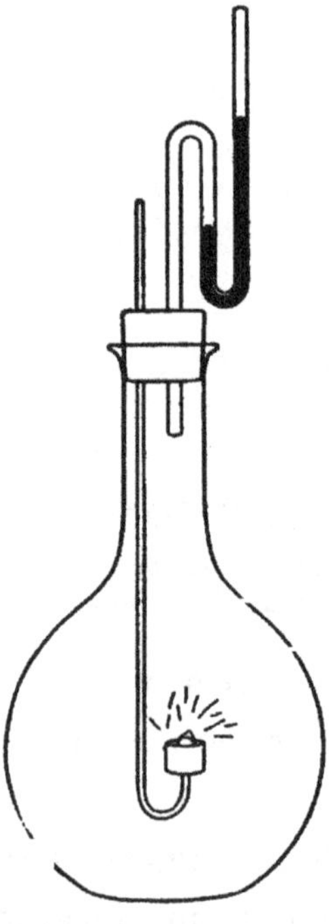

FIG. 61. Apparatus for showing volume changes which occur when sulphur and oxygen unite

Sulphur is a solid substance, and its volume is so small compared with that of the gases involved in the reaction that we can neglect it. The only volumes we need to consider are

(1) the volume of the oxygen used to burn the sulphur;

(2) the volume of sulphur dioxide produced from this volume of oxygen.

One molecule of oxygen (O_2) has been used up in the formation of one molecule of sulphur dioxide (SO_2), and since one molecular weight of any gas occupies the same volume (22.4 liters) there should be *no change of volume.* The rise of the mercury in the manometer was due to the expansion of the gases inside the flask, caused by the heat of the burning sulphur. As the gas cooled, the mercury returned to its original position.

Does the nitrogen of the air have anything to do in this experiment? Explain.

Experiment 109. Open the flask just long enough to pour in a little water, then stopper it again. Shake the water about in the flask and observe the manometer. Explain. Preserve this water for later experiments.

What are the physical properties of sulphur dioxide, — color, odor, solubility in water? Calculate its density compared with air; compared with hydrogen; compared with water.

Experiment 110. Examine the properties of the solution of sulphur dioxide in water, — color, odor, taste. Is it acid or alkaline?

146. Sulphurous acid. Since a solution of sulphur dioxide in water is an acid, it must contain hydrogen ion. The formula for the gas is SO_2, so the acid properties must be due to hydrogen from the water in which it dissolves. A reaction of the form

$$SO_2 + H_2O = H_2SO_3$$

has taken place, and a new acid, called *sulphurous* acid, has been formed. It resembles carbonic acid (Exp. 63) in that it can exist only in solution in water. If we try to concentrate the solution of sulphurous acid to obtain the pure substance H_2SO_3, the sulphur dioxide leaves the water and escapes as a gas just as carbon dioxide does.

147. Sulphites. Pure sulphurous acid is therefore unknown, but its salts, which are called *sulphites*, are stable and well-known substances which can be prepared by neutralizing the aqueous solution of sulphurous acid with the base corresponding to the salt desired. These salts, and the solution of the acid in water as well, contain the sulphite ion SO_3^{--}, which is *bivalent*, since twice the combining weight of hydrogen ion is always present with one combining weight of the sulphite ion. Sulphur dioxide and carbon dioxide are substances which are themselves not acids, but which form acids when they dissolve in water. They are called the *anhydrides* of the corresponding acids, sulphurous acid and carbonic acid.

Sulphur dioxide is a strong bleaching agent, cheaper even than chlorine and less injurious in many cases to fabrics. Silks and straw goods (but not cotton) are often bleached by this gas.

Experiment 111. Arrange a jar with a crucible filled with sulphur at the bottom. Ignite the sulphur and hang some flowers in the jar.

Shake up a little blue litmus solution in a jar containing sulphur dioxide. Try to restore the color with an acid or an alkali.

QUESTIONS

1. What are the physical properties of sulphur?
2. State what you know about the allotropic forms of sulphur.
3. How could you prove that these are all sulphur?
4. Aside from their difference in physical properties what other important difference do these allotropic forms show?
5. How does sulphur occur in nature? How is it purified?
6. State all you have learned about the reaction

$$S + O_2 = SO_2.$$

7. What are the physical properties of sulphur dioxide? What is its combining weight?
8. What are the properties of a solution of sulphur dioxide in water?
9. Tell what you can about sulphurous acid.
10. What is meant by an acid anhydride?
11. Do you think that dry sulphur dioxide would have any action on litmus paper? Explain.

PROBLEMS

1. What weight of sulphur must be burned in pure oxygen to produce 10 liters of sulphur dioxide (SO_2)?
2. What weight of sulphurous acid (H_2SO_3) can be produced from 10 liters of sulphur dioxide (H_2SO_3)?

CHAPTER XVII

SULPHURIC ACID

148. Sulphuric acid. This is a very important substance in the manufacture of the other acids and chemical products in general. It is made in quantities measured in millions of tons, especially in England and Germany. It is usually made from sulphur dioxide, water, and the oxygen of the air.

Experiment 112. Test some of the solution of Experiment 109, which has been standing for several hours, for the presence of the sulphate ion, by adding a few drops of a solution of any soluble barium salt.

The formula for sulphuric acid is H_2SO_4. Write the reaction by which it is produced from sulphurous acid.

149. Catalysis. The oxidation of sulphur dioxide or sulphurous acid by air under ordinary conditions is so slow that it cannot be used as a commercial method of making sulphuric acid. The manufacturing processes all depend on the use of a catalyser, and this is a substance which increases the velocity of the reaction by which sulphurous acid is oxidized to sulphuric, or sulphur dioxide to sulphur *trioxide.* This latter substance has the formula SO_3 (not SO_3^{--}), and with water it forms sulphuric acid after the formula

$$SO_3 + H_2O = H_2SO_4,$$

which is just like the reaction by which sulphurous acid is formed from sulphur dioxide and water. Sulphur trioxide is the anhydride of sulphuric acid.

Years ago sulphuric acid was made by distilling sulphate of iron. This substance contains water and when heated breaks up into sulphuric acid and an oxide of iron. The process was expensive and is no longer used.

The first catalyser that was used to hasten the oxidation of sulphur dioxide is still in use on a commercial scale. This is a mixture of some of the oxides of nitrogen which have the property of giving up their oxygen very readily.

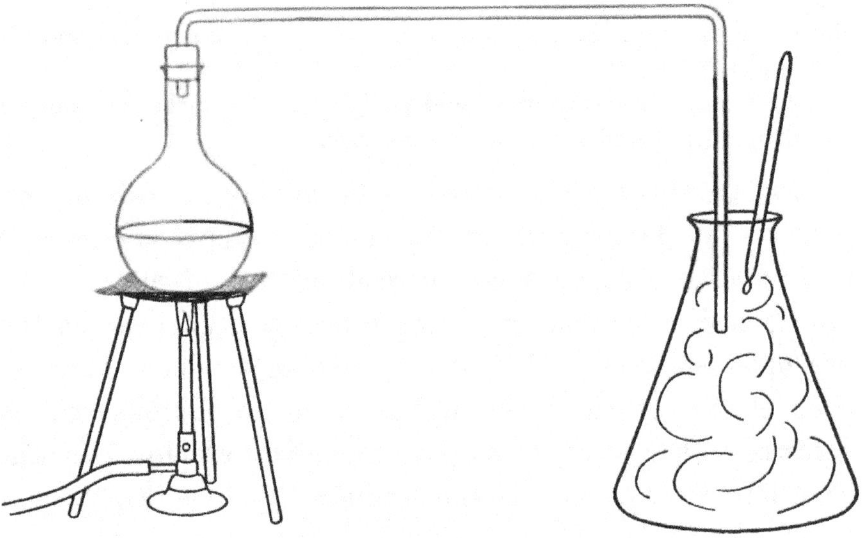

Fig. 62. Formation of sulphuric acid from sulphur dioxide, water, and oxygen

Experiment 113. Burn some sulphur in a large flask. Arrange a boiling flask with delivery tube so that steam can be blown into the large flask containing sulphur dioxide. Fasten a piece of filter paper to a glass rod, moisten it with nitric acid, and lower it into the large flask where sulphurous acid is already being formed.

Test for the sulphate ion.

150. The lead-chamber process. Experiment 113 illustrates the principle of the process by which sulphuric acid has been manufactured for many years. In the sulphuric acid works the large glass flask is replaced by huge chambers lined with lead, and the burning piece of sulphur by rows of furnaces where sulphur or iron pyrites (a sulphide of iron) is burned in a strong current of air. The mixture of sulphur dioxide and air is carried into the chambers, together with steam blown in through fine jets, and gaseous oxides of

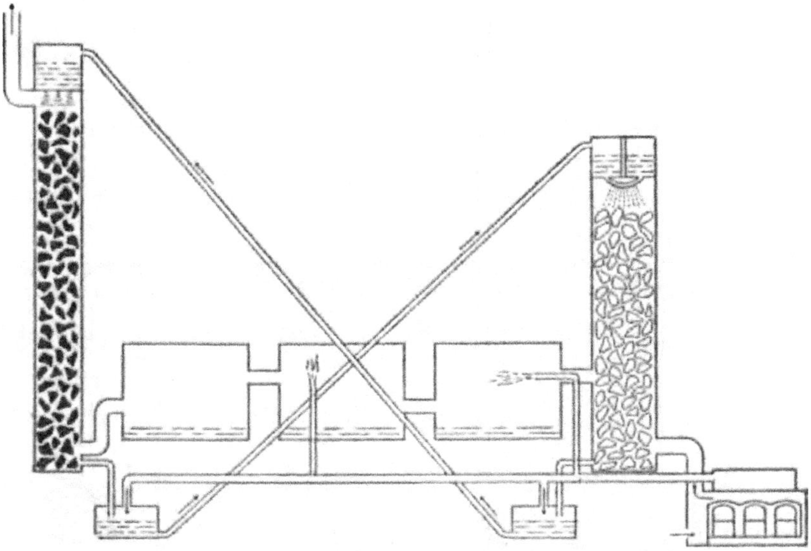

FIG. 63. General scheme of the "chamber process" for the manufacture of sulphuric acid

nitrogen, made from nitric acid, are led into the same chambers. Sulphuric acid is formed, together with the red fumes which were visible in our experiment. These red fumes are what remains of the oxides of nitrogen after they have given up a part of their oxygen to form sulphuric acid. They are led out of the chambers and brought back to their original condition by contact with air.

The final result can then be expressed by the reaction

$$2\,SO_2 + 2\,H_2O + O_2 = 2\,H_2SO_4,$$

and since the catalyser has been returned to its original condition, only sulphur dioxide, water, and oxygen from the air have been used in the process.

The lead-chamber process yields a rather dilute acid, as it goes on best in the presence of a good deal of water, and the lead of the chambers is rapidly attacked by a concentrated

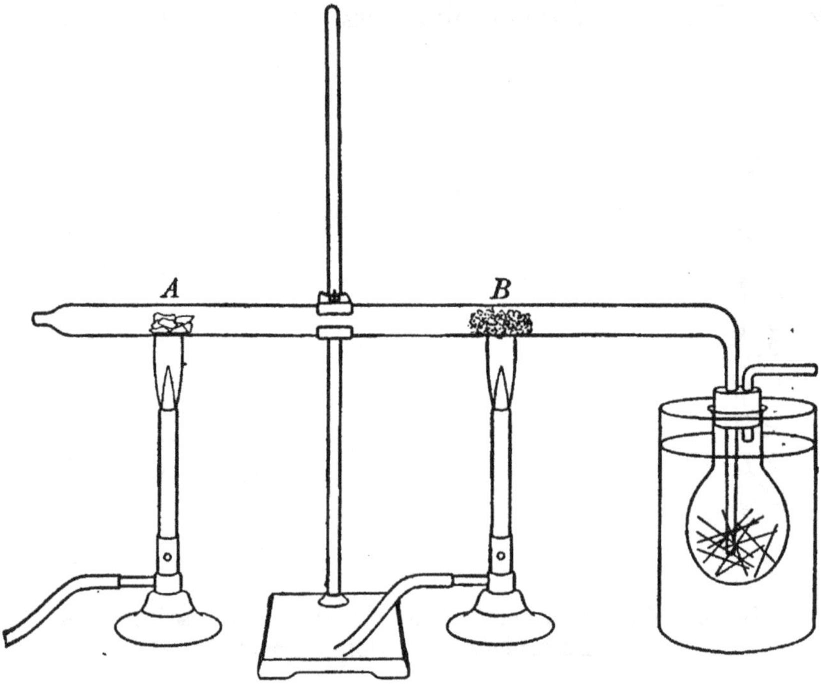

FIG. 64. The reaction of the "contact process" for the manufacture of sulphur trioxide and sulphuric acid

acid. This dilute acid is used for many purposes and its cost is very low, — six or eight dollars a ton. For many other purposes a stronger acid is required, and this is made from the "chamber acid" by concentration, first in lead and finally in platinum pans, where the water is driven off by heat.

151. The contact process. During the past three or four years another process, depending on a similar reaction but requiring a different catalyser, has become a commercial success and threatens to drive the older method from the market. This is called the *contact* process, and it depends on the fact that finely divided metals, platinum especially, have the power of making some reactions go on very much more quickly.

It differs from the older process in the fact that no water need be present. Instead of the reaction of section 150 we have

$$2\,SO_2 + O_2 = 2\,SO_3.$$

Experiment 114. Arrange an apparatus like that in Figure 64. Burn sulphur at *A* and draw the mixture of sulphur dioxide and air through the plug *B* by means of the aspirating arrangement. *B* is a plug of platinized asbestos (see Appendix, page 285).

Test the properties of the substance which is formed,—solubility, acid properties of solution, presence of sulphate ion.

Sulphur trioxide is formed, and this combines with water to form sulphuric acid (sect. 149). By the contact process sulphur trioxide is produced, and if this is dissolved in water in the proportions of 80 parts of sulphur trioxide to 18 parts of water, *pure* sulphuric acid, containing 100% H_2SO_4, results. (Why 80 parts to 18 ?)

152. Vapor pressure. Many of the uses of sulphuric acid in chemical industries depend on the fact that it has a very low vapor pressure at ordinary temperatures, and therefore a high boiling point (see Table of Boiling Points, Appendix, page 281). We have already made use of this fact in the preparation of hydrochloric acid (sect. 56). HCl is a gas, and it is, besides, only very slightly soluble in strong sulphuric acid, so it escapes from the mixture as fast as it is formed. The fact that sulphuric acid can be used in this way to drive other acids out of their salts gave rise to the idea that sulphuric acid was the strongest acid. We have learned that the acid properties of a substance depend on the presence of hydrogen ion, and that what is correctly called the strength of an acid is determined by the concentration of hydrogen ion and not by other characteristics. Sulphuric acid can be used in the preparation of the other acids, not because it is stronger, but because the others are much more volatile and escape from the reacting mixture in the form of gas before the boiling point of sulphuric acid is reached.

We have tried a rough test for the concentration of hydrogen ion in studying hydrochloric and hydrobromic acids (sect. 134). The same test will be applicable here.

Experiment 115. Prepare equivalent solutions of hydrochloric, sulphuric, and acetic acids, as described in Experiment 98. Allow them to act on metallic magnesium in apparatus similar to that of the same experiment. What is the order of strength of these three acids?

153. A drying agent. Strong sulphuric acid takes up water very readily, and it is often used as a drying agent for gases and other substances with which it does not react. It has so strong an attraction for water that it can take it away from many organic substances.

Experiment 116. Test the effect of concentrated sulphuric acid on paper, sugar, wood, and other organic substances.

The three organic substances mentioned are carbohydrates, and they contain, besides carbon, hydrogen and oxygen in the proportions in which they form water. The strong acid breaks up and destroys the organic substance by taking away water.

154. Sulphates. Sulphuric acid is a *di-basic* acid; that is, there are two combining weights of hydrogen in one combining weight of the substance, both of which go to form hydrogen ion when it is dissolved in water. It can form two series of salts, in one of which only one of the combining weights of hydrogen is replaced by a metal, while in the other series all the hydrogen is so replaced. The two sodium salts are $NaHSO_4$ and Na_2SO_4. The first of these is an *acid salt*, since a solution of it contains not only sodium ion and sulphate ion but also some hydrogen ion. It must therefore have properties corresponding to all three ions.

Experiment 117. Test the action of *acid* sodic sulphate on metallic magnesium. Has this salt acid properties? Test the solution for the presence of the sulphate ion. Hold a little of the salt on a platinum wire in the flame of a Bunsen burner. A yellow flame indicates the presence of sodium (see sect. 213).

Try the same tests with the *neutral* sodic sulphate.

The neutral salt is called Glauber's salt. It is used in medicine and to a very large extent in the manufacture of glass. We shall learn about other sulphates in the chapters on the metals, but we already know their chief characteristics as far as the sulphate ion is concerned, and what we have to learn about each individual salt are the facts about its solubility and crystalline form and the properties of the other ion which is present in its solutions.

QUESTIONS

1. From what raw materials is sulphuric acid made?

2. What need is there for a catalytic agent in the commercial manufacture of the acid?

3. Write the reactions of the contact process for making sulphuric acid.

4. What is the final reaction of the lead-chamber process (leaving out of account the reactions of the catalyser)?

5. Explain the function of the "nitrous fumes" (as the mixture of oxides of nitrogen is called) in the lead-chamber process.

6. Is sulphuric acid a stronger acid than hydrochloric? Explain.

7. Explain why sulphuric acid can be used to prepare hydrochloric acid from its salts.

8. What are the physical properties of sulphuric acid?

9. What effect has strong sulphuric acid on organic substances like sugar, paper, and wood? Explain.

10. What ions are present in a solution of acid sodium sulphate? in a solution of neutral sodic sulphate?

11. Explain how you could test for each of these ions.

12. Calculate the percentage composition of sulphur dioxide and sulphur trioxide. Show how the law of multiple proportions fits this case.

13. How many pounds of sulphur would be required to prepare 100 pounds of sulphuric acid?

14. Write the reaction which expresses the formation of
   (1) neutral sodic sulphate from common salt and sulphuric acid;
   (2) acid sodic sulphate from the same materials.

15. What is meant by a di-basic acid?

16. Write the symbols for
   Potassic acid sulphate,
   Potassic sulphate (neutral salt),
   Calcium sulphate,
   Cupric sulphate.

17. Why is sulphuric acid a good drying agent?

PROBLEMS

1. How much sulphur would be necessary to make 1000 kgm. of sulphuric acid? How much oxygen (by weight)?

2. In the reaction of section 150 how much water must be supplied for each 100 gm. of sulphur used?

3. How much sodium is there in 120.12 gm. of acid sodium sulphate?

4. What weight of Glauber's salt could be made from 100 gm. of sulphuric acid?

CHAPTER XVIII

HYDROGEN SULPHIDE

155. Reaction. Just as sulphuric acid reacts with sodic chloride (common salt) with the formation of hydrochloric acid gas, so almost any acid except the very weakest can act on many of the metallic sulphides (ferrous sulphide particularly) with the formation of a gas of formula H_2S. The reaction is

$$FeS + 2\,HCl = H_2S + FeCl_2$$

or

$$FeS + H_2SO_4 = H_2S + FeSO_4.$$

156. Properties. *Hydrogen sulphide* is the name of this gas (or "sulphuretted hydrogen"). It is somewhat poisonous, and is especially distinguished by an exceedingly bad odor (rotten eggs have the odor of hydrogen sulphide), so that all experiments with the gas should be performed either under a hood or in the open air.

Experiment 118. Arrange a generating flask like the one used in Experiment 42. Partly fill it with pieces of ferrous sulphide. Add hydrochloric acid from the dropping funnel and pass the gas produced into water. Examine its properties.

Hold a piece of polished copper over the mouth of the flask where the gas is being dissolved. Try the same with a piece of polished silver.

Hydrogen sulphide is a little heavier than air. (Calculate its density compared to hydrogen; compared to air.) It forms what appear to be salts, — ferrous sulphide, for example, —

which are compounds containing two elements, and are therefore like the chlorides, bromides, and oxides in this respect. The reaction of these salts with acids is very much like the reaction of a halogen salt with sulphuric acid. Hydrogen sulphide is an acid and ferrous sulphide is its iron salt.

Experiment 119. Connect a tip with the generating apparatus, as in Experiment 43. Is hydrogen sulphide combustible? What are the products of combustion? Hold a dry beaker inverted over the flame. Hold a cold piece of porcelain or glass in the flame. What deposits on it?

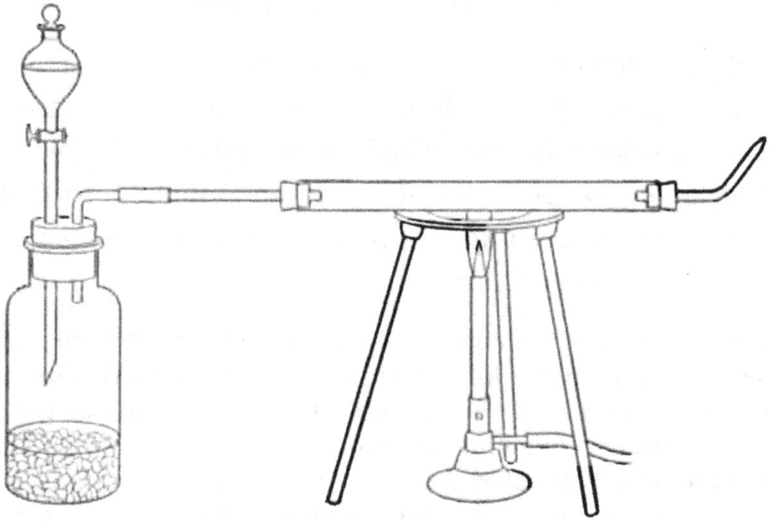

FIG. 65. Apparatus for showing the decomposition of hydrogen sulphide by heat

157. A cool flame. When the flame is cooled only the hydrogen burns, and the sulphur is cooled so fast that it does not burn, but settles on the glass just as soot does in a candle flame under the same conditions. In this latter case the carbon (of the carbon compounds in the candle) is cooled so fast that it cannot burn.

Experiment 120. Pass hydrogen sulphide gas through a tube arranged as in Figure 65. Heat the tube with a burner. Explain.

158. The oxidation of hydrogen sulphide. If we watch the solution made by passing hydrogen sulphide gas into water, it is evident that it does not remain clear many hours. The reaction

$$2\,H_2S + O_2 = 2\,H_2O + 2\,S$$

takes place, not rapidly but with measurable velocity, in a solution exposed to the air. Hydrogen sulphide gives up its hydrogen very readily indeed, and it is therefore a very good reducing agent. Its hydrogen combines sometimes with oxygen itself, as in the last equation, and sometimes with the oxygen of another substance. One other way by which it can act as a reducing agent is to give up its hydrogen to combine with other things than oxygen.

Experiment 121. Place a little iodine in a tube with some water and pass hydrogen sulphide into the water. Test the substances produced. Write the reaction.

159. Oxidation and reduction. We have here taken advantage of the reducing power of hydrogen sulphide to make hydriodic acid, sulphur being set free. This suggests a more general use of the word "reduction" than we have heretofore used. We will use this word and its opposite, "oxidation," in this broader sense:

Oxidation means not only adding oxygen but also *taking away hydrogen.*

Reduction means not only taking away oxygen, but also *adding hydrogen.*

In the equation of section 158 hydrogen sulphide is oxidized by oxygen to sulphur. Oxygen is reduced by hydrogen sulphide to water.

In Experiment 121 hydrogen sulphide is oxidized to sulphur by iodine. Iodine is reduced by hydrogen sulphide to hydriodic acid.

These same two words are also used to express similar relations and reactions in which ions enter. We say, for example, that the reaction

$$2\,I^- + Cl_2 = 2\,Cl^- + I_2$$

(see Exp. 101) represents the *oxidation* of iodide ion to free iodine, chlorine being at the same time reduced to chloride ion.

The addition of the stroke ($^-$) represents *reduction*. The removal of the stroke represents *oxidation*.

When an acid acts on metallic magnesium the reaction is

$$Mg + 2\,H^+ = Mg^{++} + H_2,$$

and we would express this by saying:

Magnesium is oxidized to magnesium ion at the same time that hydrogen ion is reduced to free hydrogen.

The addition of the sign ($^+$) represents oxidation. The removal of this sign represents reduction.

160. A weak acid. Hydrogen sulphide is a very weak acid, for its color change with litmus is only about as strong as that of carbonic acid. Its weakness is also shown by the ease with which the other acids, even the volatile ones like hydrochloric and hydrobromic acids, decompose its salts, the sulphides.

161. Sulphides. Only a few of the sulphides are soluble to any extent in water.

Solutions of sodium sulphide or potassium sulphide can be used when sulphide ion is needed.

Experiment 122. Arrange test tubes containing solutions of the following ions: zinc, cadmium, ammonium, bismuth, lead, copper, silver, sodium, calcium. Pass hydrogen sulphide gas into each of these solutions and note the color and other properties of the precipitates which are formed.

The sulphides of the different metals have such characteristic colors that they give convenient tests for the presence of the metals, and they are consequently of great aid in analytical work. Not only are the colors characteristic, but the precipitates formed are so slightly soluble that a very slight amount of a metal may often be detected by passing hydrogen sulphide gas into a solution suspected of containing it.

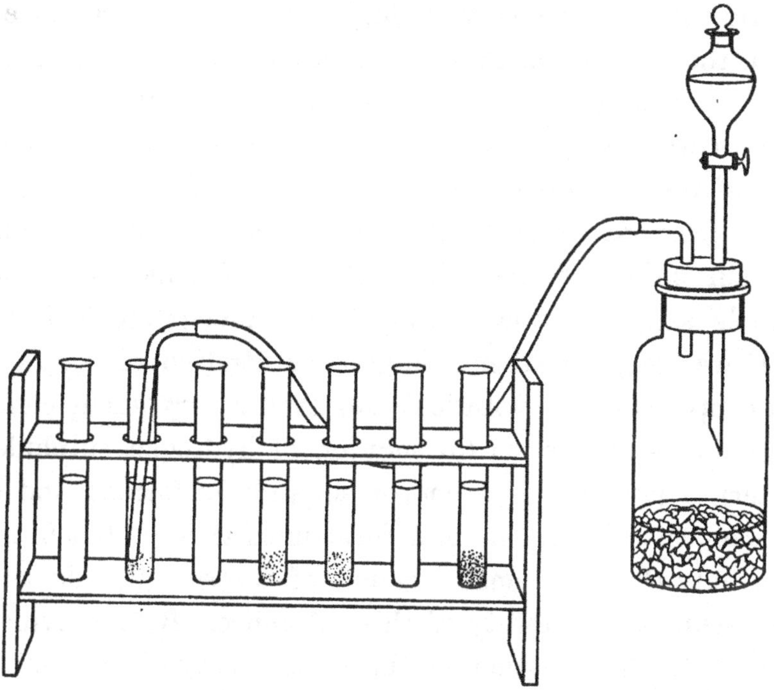

FIG. 66. Apparatus for showing the properties of some of the sulphides

For the same reason a very small amount of hydrogen sulphide, which is not even evident to the sense of smell, can be detected by adding to a solution containing it a little of a salt of a metal which forms an insoluble sulphide.

162. Reversible reactions. If we examine the reaction for the precipitation of copper sulphide from copper sulphate solution by hydrogen sulphide,

$$CuSO_4 + H_2S = H_2SO_4 + CuS,$$

it seems a contradiction to what we have already said. Sulphuric acid is produced during the reaction, and the statement has been made that sulphides are decomposed by strong acids with the evolution of hydrogen sulphide. Why does not the reaction go in the opposite direction? The reason is that cupric sulphide is a very slightly soluble substance indeed.

In these reactions in water solutions we are dealing with ions. If the substance in question is "very insoluble" (or, more correctly, only very slightly soluble), the concentration of its ions in the solution is very small. If we increase the concentration of these ions by adding solutions which contain either of them in the form of some other more soluble salt, we know already that a precipitate is formed.

If we bring together a large concentration of copper ion (in copper sulphate solution) and even a very small concentration of sulphide ion (in solution of hydrogen sulphide in water), a precipitate of copper sulphide is formed, and this goes out of solution in solid form until there is left in solution only the very small amount of copper sulphide which represents the solubility of this substance. Before giving an answer to the question of the reversibility of the reaction at the beginning of this section, let us examine some other similar cases.

Experiment 123. Make a solution of zinc sulphate in water and precipitate zinc sulphide from it as completely as possible by passing hydrogen sulphide gas through the solution for some time. Filter off some of the clear liquid and add a little caustic soda solution.

Remove some of the precipitate from the filter, mix it with water, and add a few drops of sulphuric acid.

The reaction

$$ZnSO_4 + H_2S \rightleftharpoons ZnS + H_2SO_4$$

CLAUDE LOUIS BERTHOLLET (1748–1822)

Recognized the importance of the study of chemical equilibrium and reaction velocity has been made to proceed first in one direction and then in the other. We first arranged to have in the solution zinc ion and an excess of hydrogen sulphide, together with the small amount of sulphuric acid produced during the reaction. Even this amount of acid was sufficient to keep some of the zinc sulphide in solution, and this was precipitated by neutralizing the acid.

In the second case we arranged to have present an excess of sulphuric acid, zinc sulphate (zinc ion), and only the small amount of hydrogen sulphide produced during the reaction.

Experiment 124. Pass hydrogen sulphide gas into a solution of ferrous sulphate.

Ferrous sulphide breaks up immediately in the presence of even the smallest quantity of any acid, forming ferrous ion and hydrogen sulphide. It can therefore not be precipitated in the presence of an acid, but only by neutral or alkaline sulphides.

$$FeSO_4 + K_2S = FeS + K_2SO_4$$

No acid is produced by this reaction, so a precipitate is formed.

Now hydrogen sulphide is a very weak acid indeed. This means that its solution contains only a very small concentration of hydrogen ion and a correspondingly small concentration of sulphide ion. And the addition of hydrogen ion (by adding some strong acid to the solution) decreases the concentration of the sulphide ion still more, because the reaction

$$2\,H^+ + S^{--} = H_2S$$

takes place. The stronger the acid which is added, and the greater its concentration, the less the concentration of sulphide ion which remains in solution.

Now for the answer to the question we asked at the beginning of this section. Of the three sulphides we have been studying, copper sulphide is least soluble, zinc sulphide next, and iron sulphide is most soluble of the three. The reaction

$$Cu^{++} + S^{--} = CuS$$

gives a precipitate of solid copper sulphide for very small values of Cu^{++} and S^{--} concentrations (or in very dilute solutions).

It takes a large concentration of hydrogen ion in the reaction

$$2\,H^{+} + S^{--} = H_2S$$

to make up for the very small concentration of the sulphide ion in solutions of copper sulphide. This is another way of saying that copper sulphide is soluble in concentrated strong acids but not measurably soluble in dilute acids or weak ones; or we can also say that copper sulphide will be precipitated by hydrogen sulphide from solutions which are quite strongly acid.

Zinc sulphide and iron sulphide, although what would be called "insoluble" substances, are very much more soluble than copper sulphide. A far lower concentration of hydrogen ion causes the reaction

$$2\,H^{++} + S^{--} = H_2S$$

to go on, because of the greater concentration of sulphide ion in their solutions. They are therefore decomposed by acids with evolution of hydrogen sulphide gas. Zinc sulphide is less soluble than ferrous sulphide, and it can be partially precipitated in very dilute acid solution, while ferrous sulphide can only be formed in neutral or alkaline ones.

163. Analytical separation. The reactions of the sulphides are of much importance in analytical chemistry, where it is often necessary to first get a substance into solution as a whole and then to separate the various metals in it by taking advantage of differences in the solubility of their compounds. It is very easy to separate iron and copper if they are together in the same solution. How would you do it?

It should always be remembered that even so slightly soluble a substance as copper sulphide is still soluble in some degree in water and dilute acids, so that the separation from iron will never be *absolutely* complete.

QUESTIONS

1. What are the properties of hydrogen sulphide? Is it an acid? Is it a strong or a weak one?

2. State what you know of the reducing power of hydrogen sulphide.

3. What do you understand by the word "oxidation"?

4. What do you understand by the word "reduction"?

5. State what you know of the properties of the sulphides of zinc, cadmium, bismuth, copper, silver, sodium, calcium.

6. Write the reaction for the precipitation of copper sulphide from a solution containing copper ion, by hydrogen sulphide. Explain why the reaction goes in this direction rather than in the reverse one.

7. Write the reaction which takes place when hydrogen sulphide is passed through a solution of zinc sulphate.

8. Why is no precipitate of ferrous sulphide visible when you pass hydrogen sulphide into a solution of ferrous sulphate?

9. How could you separate iron from copper if both are present in the same solution?

10. Devise a scheme for separating silver, zinc, and copper from each other.

PROBLEMS

1. What weight of ferrous sulphide will yield 10 liters of hydrogen sulphide gas?

2. What weight of hydrochloric acid will have been used up in producing this volume of H_2S?

CHAPTER XIX

NITRIC ACID

164. Saltpeter. Among the naturally occurring compounds which have been known and used for a very long time there is one called *saltpeter.* It is found in India and some other tropical countries and is the potassium salt of an acid called *nitric acid.* Long after the discovery of saltpeter, large deposits of the sodium salt of this same acid were found in Chili, and this substance was called Chili saltpeter.

165. Combined nitrogen. It has already been explained that nitrogen in combination is much more valuable than free nitrogen, and that work must be done to get it into combination, so these great beds of saltpeter are of immense commercial value because of the large amount of combined nitrogen which they contain. We do not know with certainty how the energy was supplied, nor what the reaction was which produced these deposits, but it is believed that they were produced by the action of bacteria on organic substances. The bacteria, as living organisms, drew on some store of energy — probably the oxidation of carbon to carbon dioxide — for the work necessary to cause the nitrogen to enter into combination.

The sodium salt, Chili saltpeter, is at the present time much more plentiful than the potassium salt, and it can therefore be bought much more cheaply, but it cannot be used for many important purposes because it has the property of absorbing water from the air and remaining constantly moist. Potassium nitrate (ordinary saltpeter) remains dry under the same conditions. This property of taking up moisture from the air is called *deliquescence*, and sodium nitrate is a slightly deliquescent salt. Calcium chloride and magnesium chloride, with which we are already acquainted, are very deliquescent salts.

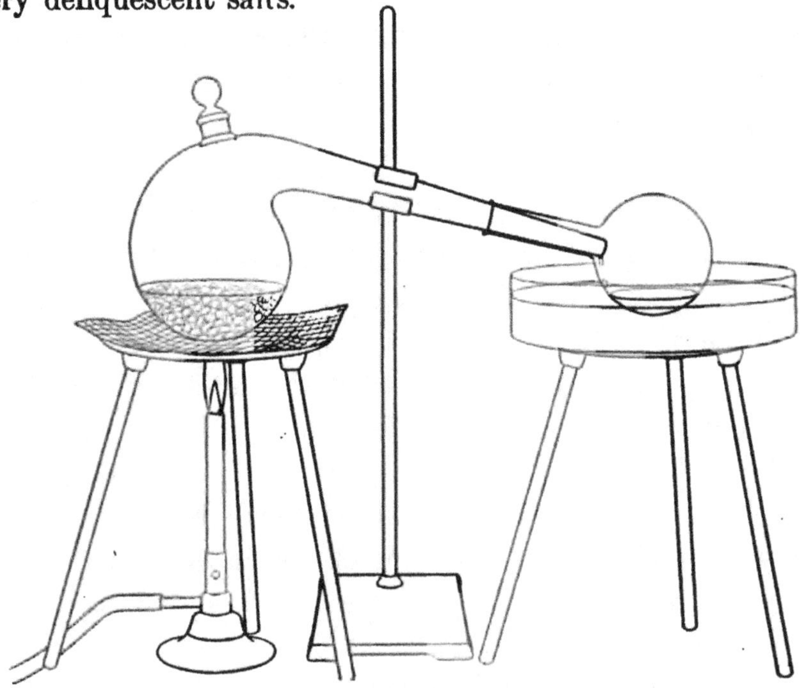

FIG. 67. Apparatus for making nitric acid

166. Nitric acid. Although sodium nitrate cannot be used for many purposes to replace the more expensive potassium salt, it can be used to equally good advantage for the manufacture of nitric acid. This process is like the general one by which a volatile acid is made, — a salt is decomposed by sulphuric acid. We take advantage of the low vapor pressure (high boiling point) of sulphuric acid, and separate from the mixture of salts and acids the most volatile substance, in this case nitric acid.

Experiment 125. Arrange apparatus as in Figure 67, using a retort partly filled with sodium or potassium nitrate, over which is poured strong sulphuric acid. The receiver is to be cooled by surrounding it with water in a basin. Examine the properties of the vapor and the liquid which is produced.

The reactions are

(1) $KNO_3 + H_2SO_4 = KHSO_4 + HNO_3$,

(2) $KNO_3 + KHSO_4 = K_2SO_4 + HNO_3$.

The first reaction takes place much more easily than the second, which requires a high temperature.

At temperatures even as low as 100°, nitric acid gives up a part of its oxygen, with the formation of oxides of nitrogen, which are yellow gases; hence the yellow vapors.

167. Properties. Pure nitric acid is a liquid, colorless and heavier than water, but under the circumstances it absorbs these yellow gases and becomes yellow itself. The smallest traces of dust and organic matter are very active in reducing the acid and causing these yellow vapors to appear. The strong acid is a very corrosive substance which produces most painful burns if it touches the skin.

Experiment 126. Test strong nitric acid with litmus paper. Test dilute nitric acid in the same way.

168. Oxidizing power. Since nitric acid gives up a part of its oxygen so readily, it is a strong oxidizing agent, and it acts very rapidly indeed when the acid is concentrated, — so rapidly that it destroys the coloring matter of the litmus paper before the usual color change from blue to red can appear. In dilute solution the oxidizing action is slow, and so the reaction for hydrogen ion takes place before the oxidation can destroy the color.

Experiment 127. With a glass rod place a drop or two of strong nitric acid on a piece of copper.

Here, as usual, an acid has acted on a metal with the formation of a salt, cupric nitrate in this case. But no hydrogen is formed because of the strong oxidizing action of nitric acid, which changes the hydrogen to water as fast as it is formed. The yellow vapors result from the reduction of the nitric acid.

Copper is not acted on at all by sulphuric or hydrochloric acids (if they are dilute) with evolution of hydrogen gas, as zinc and magnesium are. In fact, copper is *reduced* from the state of cupric ion (Cu^{++}) to the metallic state (Cu) by hydrogen gas. But in the last experiment the hydrogen is immediately removed from the reaction. It is oxidized to water and cannot show its ordinary reducing properties. Silver, which is reduced to the state of metal by hydrogen gas even more easily than copper, dissolves readily in nitric acid.

Experiment 128. Place a small piece of silver in a test tube with dilute nitric acid and apply heat.

The metal is dissolved with the formation of silver nitrate. The solution therefore contains silver ion (test for its presence).

169. Solvent power. Nearly all metals dissolve in nitric acid, even those which are easily reduced from the state of ion to the metallic state by hydrogen gas. The strong oxidizing action of nitric acid is the cause, and it acts as explained above. Only very "noble" metals (see sect. 253), like gold and platinum, are not attacked by it. Some metals, like antimony and tin, are oxidized to insoluble substances by nitric acid.

When silver dissolves in nitric acid the first reaction is

$$2\,Ag + 2\,HNO_3 = 2\,AgNO_3 + H_2,$$

hydrogen gas (H_2) being formed. Then follows immediately

$$3\,H_2 + 2\,HNO_3 = 4\,H_2O + 2\,NO.$$

These two reactions go on together, and we can write them in one formula,

$$6\,Ag + 8\,HNO_3 = 6\,AgNO_3 + 4\,H_2O + 2\,NO.$$

In order to write the second reaction we had to use $3\,H_2$, so we were obliged to multiply the first equation by 3. The final equation should be divided by 2, to make it as simple as possible. It then becomes

$$3\,Ag + 4\,HNO_3 = 3\,AgNO_3 + 2\,H_2O + NO.$$

The reaction in the case of copper is the same in principle, the difference lying in the fact that copper is a *bivalent* metal.

First $$Cu + 2\,HNO_3 = Cu(NO_3)_2 + 2\,H_2.$$

Then $$3\,H_2 + 2\,HNO_3 = 4\,H_2O + 2\,NO.$$

We need $3\,H_2$ for the second reaction, so when they are combined we have

$$3\,Cu + 8\,HNO_3 = 3\,Cu(NO_3)_2 + 4\,H_2O + 2\,NO.$$

Experiment 129. Ignite a piece of charcoal and dip it into strong nitric acid.

This is another example of the strong oxidizing power of nitric acid. It can give up so much oxygen that it supports combustion.

170. The nitrates. Nearly all the nitrates are easily soluble in water. This fact, together with what we have already learned about the oxidizing power of nitric acid, accounts for its great solvent power.

171. Uses of nitric acid. We have already seen how useful nitric acid is in the lead-chamber process for making sulphuric acid, since it supplies the oxides of nitrogen, which act as the catalyser for that important process. It is also very useful in many other chemical processes because of its strong oxidizing power. Its salts are used in explosives, and it is a necessary substance for almost every chemical manufactûre. Its very greatest importance is in the manufacture of fertilizers, for all plants need nitrogen for their growth, and especially for their fruits and flowers. After soil has been in continual use for years it becomes exhausted, and the supply of combined nitrogen must be renewed in some way. A few plants (beans and other members of the same family) have the power of producing more combined nitrogen than they need for their own lives, making it from the nitrogen of the air with the aid of bacteria which gather in and about their roots. Soil which has been nearly exhausted of its combined nitrogen can be in some degree restored by a proper cycle of crops, including one of these nitrogen-producing plants.

172. Fertilizers. It is often best to restore exhausted soil by applying a fertilizer, and for the nitrogen portion of these mixtures we are at the present time almost entirely dependent on the Chilian beds of sodium nitrate. When these beds are exhausted, as they will be in the not very distant future, we shall be obliged to find some other source for the combined nitrogen which is so essential to agricultural success. Perhaps by that time we shall be able to use the immense store of nitrogen which is in the air about us. This we can, of course, make use of if we can find a cheap way of getting it into the form of combined nitrogen, so that it can be taken up by plants. There are several ways of doing this, but they have so far proven too expensive to compete with the still plentiful supply of sodium nitrate from South America.

173. The fixation of atmospheric nitrogen. Nitric acid is produced when an electric spark is passed through a mixture of oxygen, nitrogen, and water vapor, and this reaction is now being worked on a commercial scale, making use of great quantities of electrical energy to cause combination to take place. Where the cost of power is very small indeed, and where no other raw materials offer larger profits, this may prove a commercial success. So far the only source of power which is cheap enough is water power, and the only successful plants are in Norway, where immense falls offer a source of energy.

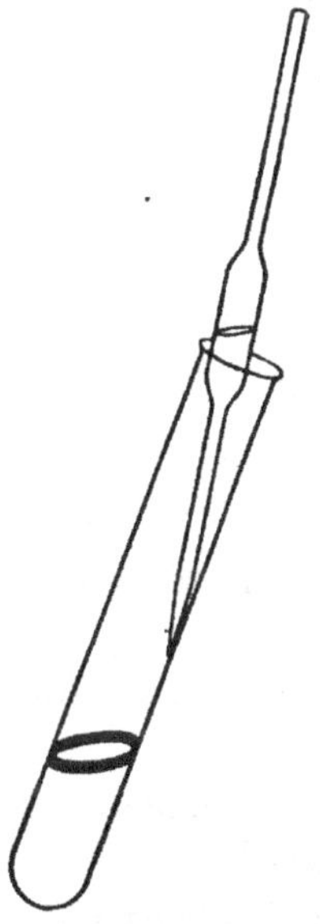

FIG. 68. Testing for nitric acid

174. Aqua regia. A striking example of the oxidizing power of nitric acid is seen in its action on hydrochloric acid. When strong nitric and strong hydrochloric acids are mixed in the proportions 3 of HCl to 1 of HNO_3, the hydrogen of the hydrochloric acid is oxidized to water by the nitric acid. Chlorine and another gas called nitrosyl chloride are set free, and these dissolve in the mixture of the two acids and turn it yellow. This mixture dissolves metals that are not soluble in either hydrochloric or nitric acid, and even gold and platinum are attacked by *aqua regia*, as this mixture is called, not because it is a stronger oxidizing agent than nitric acid (it is not), but for another reason (see sect. 302).

175. A test for nitric acid. When NO gas is passed into a solution of ferrous ion a dark liquid is formed, which contains a new complex kation made up of iron, nitrogen, and oxygen. This dark color can be used as a test for nitric acid and the oxygen compounds of nitrogen.

Experiment 180. Add a little strong sulphuric acid to a ferrous ion solution, allowing the heavier acid to go to the bottom of the tube and being careful not to shake it. Add a few drops of a solution of a nitrate (see Fig. 68).

QUESTIONS

1. What is saltpeter? What is Chili saltpeter?
2. Why are these very valuable substances?
3. What is meant by deliquescence?
4. How is nitric acid made?
5. What are the properties of nitric acid?
6. Why are copper and silver soluble in nitric and not in hydrochloric acid, which we know to be just as strong an acid?
7. Why is it necessary to supply the soil with nitrogen in the form of a fertilizer?
8. How can soil which has been exhausted of nitrogen compounds be renewed by a cycle of crops? What crops should be used?
9. What chemical test for nitric acid and the oxides of nitrogen have you used?
10. What is meant by the fixation of atmospheric nitrogen?

PROBLEMS

1. What weight of sodium nitrate must be used to make 1000 gm. of nitric acid? What weight of potassium nitrate?
2. What weights of nitric acid and metallic silver will yield 100 gm. of silver nitrate?
3. What volume of NO will be produced?
4. How much cupric nitrate can be made from 10 gm. of copper?

CHAPTER XX

AMMONIA

176. A nitrogen compound. *Ammonia* is a nitrogen compound of great importance in chemical processes in the laboratory and the factory, and in agriculture as well. It is one of the nitrogen compounds from which plants can very easily take their supply of nitrogen, and it is used to nearly the same extent as the nitrates in the manufacture of fertilizers.

Ammonia itself is a gas which is very soluble in water, and the same name is very commonly used for the aqueous solution. It is formed by the decomposition of many nitrogen-containing organic substances, and its odor is a familiar one about stables and barns. In fact, the first discovery of ammonia was made by collecting the white, saltlike substance which separates out when the drainage from stables dries in the sun.

177. Sources. At the present time the principal source of ammonia is the gas liquor from gas works and coke furnaces. The bituminous coals which are used in the manufacture of illuminating gas and coke contain large amounts of nitrogen and hydrogen, and at the high temperatures of the gas retort and the coke furnace these two elements combine to form ammonia (NH_3). This escapes with the other gaseous products of distillation and is dissolved in the water which is used to purify the gas. From this first impure solution it is driven out by distillation with lime, and the purified gas is either compressed to a liquid in steel cylinders and sold in this form, or else combined with an acid to form a *salt of ammonium.*

Experiment 131. Mix a little saltpeter with about ten times as much fine iron filings and heat the mixture gently in a test tube. Test the gas which is formed.

Experiment 132. Mix some broken pieces of caustic soda with iron filings and heat in a test tube as above. What gas is formed?

Experiment 133. Mix about 1 part of saltpeter, 2 of sodium hydroxide, and 20 of fine iron filings. Heat the mixture in a hard-glass tube. What gas is formed? Test its properties. Hold a wet piece of litmus paper in the tube.

There is evidently a base present, for the paper shows the reaction for the OH^- ion. Since ammonia itself consists only of nitrogen and hydrogen, the O must have come from the water on the damp paper. A solution of ammonia in water contains the base NH_4OH, and therefore the ions NH_4^+ and OH^-.

178. Properties. The formula for ammonia gas is NH_3. (Calculate its density compared with hydrogen; compared with air.) The most usual method of preparing it in the laboratory is to heat a salt of ammonium with an alkali, and lime being the cheapest alkali, it is used for this purpose.

Experiment 134. Mix about equal parts of ammonium chloride and freshly slaked lime, adding a very little water. Place the mixture in a flask with stopper and delivery tube passing into an absorption bottle, as in Figure 42.

The reaction is

$$2\,NH_4Cl + Ca(OH)_2 = CaCl_2 + 2\,NH_3 + 2\,H_2O.$$

The gas escapes from the mixture and may be either collected as a gas or dissolved in water.

Experiment 135. Collect some of the gas by upward displacement (see Fig. 69), and test its properties in every way you can.

Ammonia gas is very much lighter than air. It has a pungent odor very easy to recognize, and it is very soluble in water, one liter of water at ordinary temperatures dissolving about 700 liters of the gas. This water solution has all

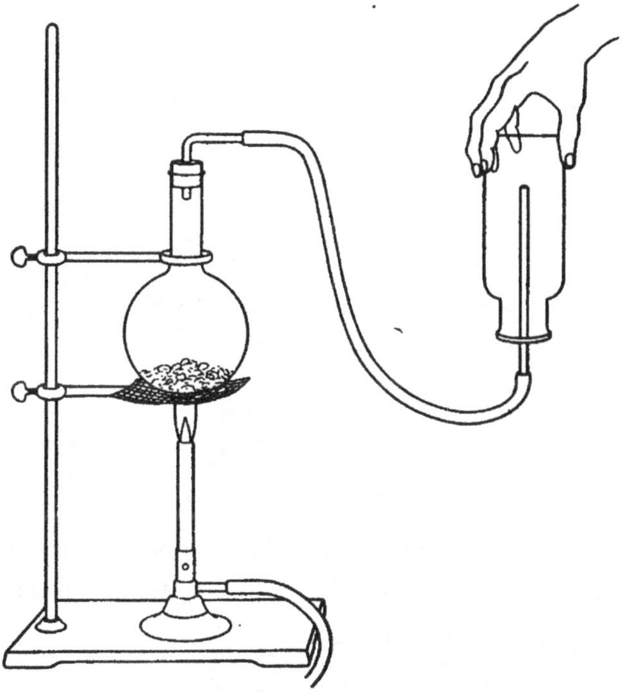

FIG. 69. Collecting ammonia gas by upward displacement

the properties of a base, and gaseous ammonia is an anhydride in the same sense that carbon dioxide and sulphur dioxide are, except that in one case the compound with water is an acid and in the other a base.

179. Ammonium hydrate. The base NH_4OH is called ammonium hydrate. It forms salts with all known acids, and these are called ammonium salts, the group NH_4

(ammonium) acting in every way like a metal except that it cannot be prepared in the free state either by chemical reactions or by electrolysis, or, so far as we know, in any other way.

180. Ammonium salts. Although ammonium cannot be prepared in the free state, its salts are cheap, common, and very useful substances. They may be prepared in two ways: either by allowing gaseous ammonia to pass into a dilute solution of an acid, in which case the reaction is

$$NH_3 + HCl = NH_4Cl \text{ (for hydrochloric acid);}$$

or by neutralizing ammonium hydroxide solution with a dilute solution of the acid, when the reaction

$$NH_4OH + HCl = NH_4Cl + H_2O$$

takes place. If we use sulphuric instead of hydrochloric acid, the reactions for the two cases will be

$$2\,NH_3 + H_2SO_4 = (NH_4)_2SO_4$$

and $$2\,NH_4OH + H_2SO_4 = (NH_4)_2SO_4 + 2\,H_2O.$$

181. General ways of making salts. There are several general ways of making salts. Take potassium and hydrochloric acid, for example. The metal decomposes the acid very rapidly, and the reaction is

$$2\,K + 2\,HCl = 2\,KCl + H_2.$$

If we use the oxide of potassium,

$$K_2O + 2\,HCl = 2\,KCl + H_2O;$$

and if we use the base, potassic hydroxide,

$$KOH + HCl = KCl + H_2O.$$

Experiment 136. Bring the mouth of the bottle containing ammonium hydroxide solution close to the mouth of the bottle containing hydrochloric acid solution.

In this case ammonia gas combines directly with gaseous hydrochloric acid to form solid ammonium chloride. The solid salt, which appears under these circumstances as a white cloud, is the same substance as ordinary ammonium chloride, often called sal ammoniac.

Experiment 137. Try the same experiment with the sulphuric acid bottle.

182. Volatile alkali. Instead of a white cloud a solid crust will form after a time around the mouth of the sulphuric acid bottle. The volatile ammonia escapes from its bottle and diffuses through the air to the other bottle, neutralizing the acid which is clinging to its lip, with formation of ammonium sulphate. The same process will take place when ammonia escapes near bottles containing other nonvolatile acids, so when ammonia and acids are kept in the same cupboard the lips of all the acid bottles will finally become crusted with solid ammonium salts, corresponding in each case with the acid in the bottle. If hydrochloric or any other volatile acid is present, the mouth of the ammonia bottle will become crusted with the corresponding salt.

183. Catalysis. Ammonia is easily oxidized to nitric acid in the presence of air and moisture.

Experiment 138. Wind a piece of rather fine platinum wire into the form of a spiral and heat it to glowing in the burner. Then lower it quickly, before it can cool, into the mouth of a bottle partly full of strong ammonia water.

The fumes which are formed consist largely of ammonium nitrate, produced by the combination of the nitric acid which is formed with the excess of ammonia gas in the bottle. The reaction is

$$NH_3 + 2\,O_2 = HNO_3 + H_2O.$$

QUESTIONS

1. What are the properties of ammonia gas?
2. What are the principal sources of ammonia?
3. What base is present in a solution of ammonia gas in water?
4. The formula for ammonia gas is NH_3. What is its density compared with hydrogen? compared with air?
5. In what respects is the NH_4 group like a metal?
6. How is ammonia gas prepared in the laboratory?
7. How are the salts of ammonium made?
8. State all the different ways of preparing ammonium salts.
9. How can nitric acid be made directly from ammonia?

PROBLEMS

1. What volume of ammonia gas can be obtained from 10 kgm. of ammonium chloride?
2. What weight of water will be produced by this reaction?
3. What weight of nitric acid can be produced from 100 gm. of ammonia? (See sect. 183.)
4. If 10 liters of ammonia gas is passed into sulphuric acid, what weight of ammonium sulphate will be produced?

CHAPTER XXI

PHOSPHORUS

184. Historical. This very interesting element was discovered long ago, in the days of alchemy, by an old philosopher named Brandt. He was searching for the philosopher's stone, — a substance which would turn everything into gold. This substance which he discovered has such strange properties that he kept the method of its preparation a profound secret. It was not until matches began to be manufactured, as late as 1833, that this element became at all common or cheap.

It is a very usual thing that whenever a substance which is not really rare (in the sense that its compounds form a very small part indeed of the earth's crust) can be applied to technical and commercial uses, it immediately becomes cheap, the price being measured only by the outlay of energy necessary to prepare it from the substances in which it occurs in nature, and also in some degree by its abundance. If a substance has only scientific use, in other words, if it is used only in the laboratory and has no technical applications, these two factors are no measure of the price of a substance.

For example, metallic sodium was an extremely expensive substance for many years after it was first made, but modern electrolytic methods have reduced it in price to thirty-five or forty cents a pound. The price of phosphorus at the present time is not more than forty or fifty cents a pound.

185. Properties. Phosphorus is a white, waxy-looking solid, which melts at 44°. It combines very readily with oxygen, so readily that it must always be handled and cut under water, and the friction of a knife against dry phosphorus is quite sufficient to set it on fire. The very greatest care is necessary in handling it, for the burns which it produces are dangerous, not only because they are very deep but also because of the poisoning they produce. Phosphorus should never be touched with the fingers under any circumstances, but always with forceps, and it should be remembered that every small bit of phosphorus which is left lying about will take fire in a few minutes or hours spontaneously.

186. Supercooling. Although the element melts at 44° and solidifies at the same temperature, it can be cooled far below this point if no solid phosphorus is present.

Experiment 139. Cut off a piece of phosphorus as big as a pea (under water!) and place it, covered with water, in a test tube. Heat gently until the phosphorus melts, and then cool the whole tube by placing it in water. At what temperature does the phosphorus solidify? After it has become cold touch a glass rod to a piece of solid phosphorus and touch the surface of the molten substance with the rod.

Just as water can be cooled far beyond its freezing point (see Exps. 33 and 34), so phosphorus can be cooled below its point of solidification, provided the solid phase is not present.

187. Allotropic forms. The element phosphorus exists in two allotropic forms, one of which can be easily transformed into the other.

Experiment 140. Dry a test tube and a small piece of phosphorus (carefully!) with filter paper. Place the phosphorus in the test tube and stopper the mouth loosely with a plug of cotton wool. Heat the test tube with a burner, gently at first and then more strongly.

The phosphorus combines first of all with all the oxygen in the tube, and after this the diffusion of air from without through the porous stopper takes place very slowly, and the phosphorus no longer burns. The substance which is slowly formed by heating ordinary white phosphorus out of contact with air is an allotropic form of the same element. It is called *red* phosphorus, and it can be easily transformed into ordinary phosphorus again.

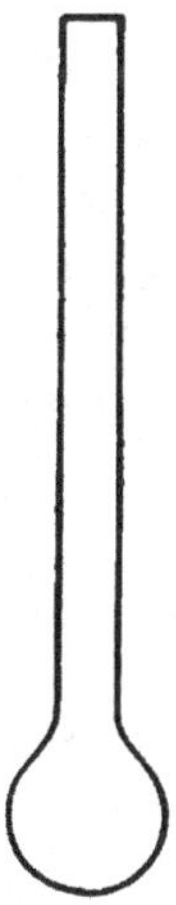

Experiment 141. Heat a little red phosphorus in a hard-glass tube stoppered as before with a plug of cotton wool. What collects in the cooler part of the tube?

188. Unstable forms. It is remarkable that by simple heating, white phosphorus should be changed into red, and red into white. It is an example of a rather general fact: of all the allotropic forms of an element one is always the most stable at any given temperature. Red phosphorus is more stable than white at all temperatures up to the boiling point, and white does, in fact, change into red even at ordinary temperatures, although the change is so slow as to be practically unmeasurable.

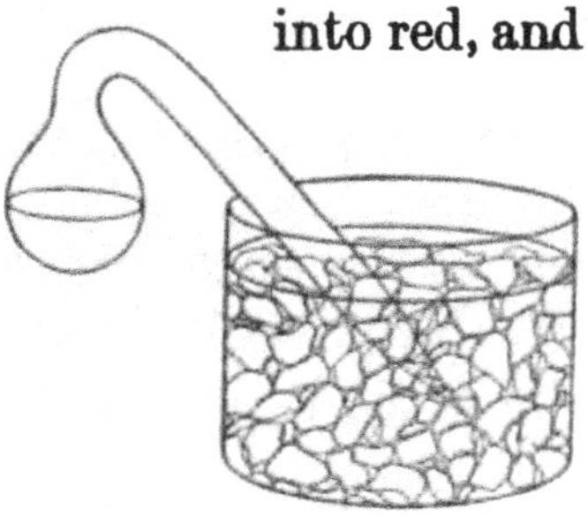

FIG. 70. Small glass retort for showing supercooling

Phosphorus vapor is the same thing, no matter which form it comes from, and if this is cooled quickly the unstable white form is first formed. This changes into red very gradually at ordinary temperatures, but with measurable velocity at high temperatures.

This may be stated more generally by saying that when a substance exists in several forms one of them is always more stable than the others at any given temperature, and that when the substance forms from another phase the less stable forms often appear first, changing slowly or rapidly, according to the temperature, into the stable one.

Experiment 142. Blow for yourself a small retort from glass tubing, making it the shape shown in Figure 70. Place some water in the bulb, draw out the other end to a point, and then boil the water in the bulb. After it has boiled for several minutes and has driven all the air out of the tube, seal off the end of the fine tube with the burner, taking care not to allow the boiling to stop long enough for air to enter the tube. The retort now contains only water and water vapor. Dip the fine point into a freezing mixture of temperature about – 5°. Water will distill from the warm bulb and condense in the cold point of the tube.

Which should be more stable at – 5°, water or ice? Which forms under these conditions?

Experiment 143. Examine the physical properties of the two forms of phosphorus. Try their solubility in carbon bisulphide. Test their ignition temperature on a metal plate over the burner, as in Figure 71, heating the two heaps equally. Which ignites first?

The following table shows how the two forms differ:

White	*Red*
Melts at 44°.	Does not melt.
Very poisonous.	Not poisonous.
Low ignition point.	High ignition point.
Soluble in carbon bisulphide.	Not soluble.
Phosphorescent in air.	Not phosphorescent.

While the physical properties of the two forms are very different, we can be sure that both are the same element, for the same substances are formed from each when they enter into combination. As in all other cases of allotropism, one form reacts with the production of more heat than the other. In this case it is the white form which gives the greatest amount of heat during its reactions, and it is in general the less stable form which does so.

189. Luminescence. Phosphorus shows some properties different from those of any element we have yet studied about. It glows in the dark; but it gives out this light only when in contact with air, and it does in fact burn, combining with the oxygen of the air to form an oxide of phosphorus. The strange thing about it is that the temperature remains low, even though light is given out. A part of the energy which is given out when phosphorus combines with oxygen is in this case given out directly as light.

FIG. 71. Showing difference in the ignition points of red and white phosphorus

When we see a body giving out light we usually think it is hot, and most bodies must be heated up to 600° or higher before they begin to glow. But this is only one of several classes of light emission. It is called *incandescence*. The glowing of phosphorus is another kind of light emission, in which the color of the light from the body has no relation whatever to the temperature. Glowworms and lightning bugs have this power of emitting light without heat, and so have the many phosphorescent animals of the sea, and some bacteria. These other forms of light emission, where the color of the light has no relation to the temperature of the emitting body, are called *luminescence*.

190. Phosphoric acid. This acid has the formula H_3PO_4. All of its three hydrogens become ions in solution, so it can form three classes of salts, as one, two, or three of the hydrogen ions are replaced by a metal.

Experiment 144. Place a small piece of phosphorus on a crucible lid and support it on a small tripod on an earthenware plate. Ignite

the phosphorus and cover it with a large inverted beaker. Collect the substance formed after it has settled, and dissolve it in a little water. Test its properties. Is it an acid?

The white substance is phosphorus pentoxide (P_2O_5). It is an acid anhydride, and when it dissolves in water, phosphoric acid is formed:

$$P_2O_5 + 3\,H_2O = 2\,H_3PO_4.$$

191. Phosphates. The bones of the animal body consist of organic substances and calcium phosphate, and when bones are burned the organic substances are driven off, and bone ash, which is largely calcium phosphate, remains behind. Phosphates are also necessary constituents of other parts of the animal body and of all plants, so the phosphates which have been removed from the soil by successive crops must be replaced. Phosphates are therefore one of the constituents of fertilizers, together with combined nitrogen and potassium salts of some kind. These three things supply nearly all the needs of plants which cannot be taken from ordinary soil and the atmosphere.

QUESTIONS

1. What are the properties of phosphorus?
2. Tell what you can about the allotropic forms of the element.
3. Explain how it is that white phosphorus can be changed into red, and red into white, by means of heat.
4. Describe the phenomena of supercooling.
5. What is the difference between incandescence and luminescence?
6. How is phosphorus pentoxide made?
7. What is produced when phosphorus pentoxide dissolves in water?
8. How many different kinds of salts can phosphoric acid form? Can you give an example of each?

PROBLEMS

1. How much phosphorus pentoxide can be made from 10 gm. of phosphorus?
2. What volume of oxygen will be required to combine with 10 gm. of phosphorus to form phosphorus pentoxide?

CHAPTER XXII

SILICON

192. Occurrence. Silicon does not occur free in nature because it combines very readily indeed with oxygen, but its compounds are among the most abundant and widespread, and in the form of these compounds it makes up about a fourth of the whole crust of the earth. The element itself is hard to obtain because of this very great affinity for oxygen. It can be produced in at least two allotropic forms, one of which is a powder, analogous to amorphous carbon, and the other a crystalline substance.

FIG. 72. Groups of quartz crystals

193. Silica. The oxide of silicon has the formula SiO_2. It is called silicon dioxide, or much more commonly *silica.* This oxide is the most common compound of silicon; sand, sandstone, and other substances which occur in enormous quantities in the earth consist of silica. Its purest form is rock crystal, or *quartz,* which occurs in fine, clear crystals, sometimes colored violet, red, or brown by impurities. These colored quartzes are called amethyst, rose quartz, and smoky quartz, and their beautiful crystals have very probably been deposited from aqueous solutions of silicic acid. They seem to have been deposited from solution rather than from fusion, because the color of most of these crystals is completely destroyed when the crystals are heated.

Other forms of silica which are imperfectly crystalline or amorphous are onyx and flint. Opal is silica combined with water, — a hydrated silica. Petrified wood is a variety of silica, usually colored, which has gradually replaced the wood fiber of a tree, changing it to stone.

194. Silicic acid. Silica is SiO_2. It is the anhydride of an acid of formula H_2SiO_3 ($SiO_2 + H_2O$), but this acid cannot be formed by dissolving SiO_2 in water, and it is similar to carbonic and sulphurous acids only in the form of its symbol and its great weakness. The acid is only known from its salts, of which a great number occur in nature, especially the salts of potassium, sodium, calcium, magnesium, aluminium, and iron. These *silicates* form the greater part of the solid crust of the earth. Most of them are insoluble in water, but the alkali silicates are soluble, and their solutions seem to contain the ion SiO_3^{--}.

The pure substance H_2SiO_3 does not seem to have ever been prepared, and its solution in water has a series of properties entirely different from those of any acid we have examined.

Experiment 145. Add hydrochloric acid slowly to a solution of sodium silicate (water glass) stirring constantly.

Water glass is made by fusing together silica and sodium carbonate. The volatile carbonic acid is driven off. (Write the reaction.)

If the sodium silicate solution is a very dilute one, no silicic acid separates out, but when the dilute solution is evaporated, the jellylike acid separates and does not go back into solution when water is added.

195. Colloids. This is the first substance we have met with which has these strange properties. There are many other such substances, and they are called *colloids.* Chemists are only just beginning to study their properties carefully, so we know very little about them.

Silicic acid is a very weak acid indeed, — so weak that its salts are broken up into acid and base in water solutions.

Experiment 146. Test water-glass solution with litmus. Is it acid or basic? What ion is present in the solution?

Add hydrochloric acid carefully until the blue color just changes to red. What changes take place in the properties of the solution?

196. Weathering of rocks. That it is an acid is shown by the fact that it forms salts, and this is, after all, the final test for an acid. But the acid is so weak that these salts, which make up so many rocks, are attacked by the carbon dioxide of the air and changed into other substances wherever the atmosphere gets at them. During this reaction the rocks crumble to pieces because the substances which bind them together are attacked, and the process is called *weathering.* It is going on continually all over the earth's surface: mountains are being broken up in this way, and brooks and streams carry the fine particles which are formed down into the valleys as sand and silt. The rivers carry them still farther and finally deposit them in the sea, where they sink to the bottom, forming layers which harden again into stratified rocks. During the process of weathering the silicic acid which is set free is in part dissolved again in the water, and it is deposited again partly as quartz and partly in the various amorphous forms of silica.

197. Natural chemical processes. The earth's crust is one great laboratory in which these processes as well as the life processes of plants and animals are being continually carried on. It is, however, easier to see and understand the processes of plant and animal life, because these go on fairly rapidly, while the weathering of mountains and the gradual leveling of the earth's surface are exceedingly slow, — so slow as to produce hardly a visible change during a man's lifetime, but none the less evident on close examination, and none the less the cause of most of the changes which are going on in the earth's surface.

198. Glass. Glass is a mixture of silicates, one of which is a silicate of potassium or sodium. Ordinary window glass is composed of the silicates of sodium and calcium, and it is made by melting together sand with a sodium compound and a calcium compound (see sect. 237). Other substances are often added to give the glass special properties, such as color.

Glass has very peculiar properties. It has no real melting point, as most ordinary solids have, but changes slowly from the solid to the liquid state, passing through a series of intermediate properties on the way. If a glass rod is heated over the burner or in the blast lamp, it first becomes soft and can be easily bent, and if it is kept in this condition it will run slowly, like thick molasses or glue. As it is heated hotter and hotter it becomes softer, and finally it can be poured just as a true liquid can. When it is cooled down again there is no sudden point of solidification where the solid and liquid phases can exist together, but only a gradual change of properties in the reverse direction. These are the properties of *amorphous* substances. Crystalline ones usually have definite melting points.

QUESTIONS

1. How does silicon occur in nature?
2. Why is the element itself difficult to prepare?
3. What is the commonest compound of silicon? What is its formula? Of what acid is it the anhydride?
4. What minerals do you know which consist largely of silica?
5. State what you know about the properties of silicic acid.
6. What are colloids?
7. What evidence have you concerning the strength of silicic acid?
8. What property decides finally whether or not a substance is an acid?
9. What is water glass?
10. Explain the process of weathering of rocks.
11. What is glass?
12. What peculiar properties does glass show?

CHAPTER XXIII

SODIUM

199. Properties of metals. There is not a very sharp dividing line between metals and nonmetals. If there is any question as to whether or not a substance is a metal, the following tests may be applied:

1. Metals are, with one exception (mercury), solids at ordinary temperatures, and most of them have high specific gravities.
2. They have what is called metallic luster.
3. They usually conduct heat and electricity much better than the nonmetals.
4. They are usually malleable and ductile.
5. They form alloys with each other.
6. They form bases. When the bases are dissolved in water the metal and the hydroxide group with which it is combined take on ionic form, so that their solution contains the ions M^+ (or M^{++}, etc.) and OH^-.
7. They combine with acids to form salts. These salts in solution contain the kation corresponding to the metal (M^+, M^{++}, etc.) and the anion corresponding to the acid (A^-, A^{--}, etc.).

These last two are the most important properties of metals. They may be combined in the statement that metals form kations, each of which has its own peculiar properties, and all of which move toward the kathode during electrolysis.

200. Sodium and potassium. Sodium and potassium are metals whose compounds are used more frequently in chemistry than those of any other metal. These metals occur in the combined state in enormous quantities on the earth's surface, — potassium compounds in larger amount than sodium, but harder to get out and use in the preparation of pure substances. As a result, compounds of potassium are a little more expensive than the corresponding sodium ones, but this difference only makes itself felt when large quantities are in question.

201. Sodium chloride. This substance occurs in immense quantities in the sea and in salt lakes, and also in great beds as rock salt in many parts of the world. In America we have no very large salt beds, and we are therefore dependent on salt from the ocean and from salt springs and lakes for our supply. But in Europe much the larger part of all sodium compounds is made from crude rock salt taken from salt beds. Where rock salt is plentiful it is cheaper to mine the impure substance and purify it by recrystallization than it is to furnish the heat necessary to evaporate the large quantity of water in which the salt of the ocean is dissolved, and also cheaper than it is to provide the large space necessary to allow the sun's heat to perform the evaporation.

The salt of the sea is very largely sodium chloride, but if it is obtained by complete evaporation of sea water it is very bitter, owing to the presence of magnesium salts. It is purified by recrystallization before it can be used for the table. The magnesium salts are much more soluble than the sodium chloride, and the latter separates in comparatively pure condition when the solution is evaporated nearly, but not quite, to dryness. Another point which aids in this process of purification is the fact that magnesium chloride and the other impurities are much more soluble in hot water than in cold, while the solubility of sodium chloride changes only very slightly with a large change in temperature (see curve, Fig. 78). Sodium chloride therefore separates from a hot concentrated solution nearly as completely as from a cold one, while the other substances remain in solution at the higher temperature and thus assist in the separation.

The sodium salts of a very great number of acids are used in the laboratory, but only a few are of great importance for technical purposes. The two carbonates of sodium, Na_2CO_3 and $NaHCO_3$, are made in enormous quantities and used for many purposes, and the sulphate of sodium is used in the manufacture of glass.

202. Metallic sodium. Since this element combines so readily with oxygen, with the evolution of a great deal of heat, it requires the expenditure of energy to obtain the free metal from its compounds. The process is a *reduction* to the metallic state, and we naturally think of carbon in the form of coal as a cheap and powerful reducing agent because it combines so readily with oxygen, taking it away from many oxides and other compounds and leaving the metal free. The older process for making sodium depended on the reducing action of coal, but a very high temperature is necessary and the metal is hard to produce. At the present time a large part of the world's supply of metallic sodium, which is used in many chemical processes as a strong reducing agent, is made by the electrolysis of molten caustic soda. Instead of making use of the energy of the coal directly for the reduction, it is found to be more economical to burn the coal and produce electrical energy, and of course cheaper still, if the electrical energy can be produced from water power. In the Castner process, which is carried out on a large scale at Niagara Falls, molten sodium hydroxide, which is made by electrolysis of sodium chloride solution, is electrolyzed in large iron pots. A diaphragm separates the electrodes, as in Figure 73, so the sodium rises to the top of the kathode compartment and is removed there. The great advantage of the electrolytic process lies in the fact that the energy can be applied where we want it and in concentrated form. The reducing action of the kathode is a very strong one, and it is used in very many important processes to replace the reducing action of coal. Later we shall find other applications of the same facts.

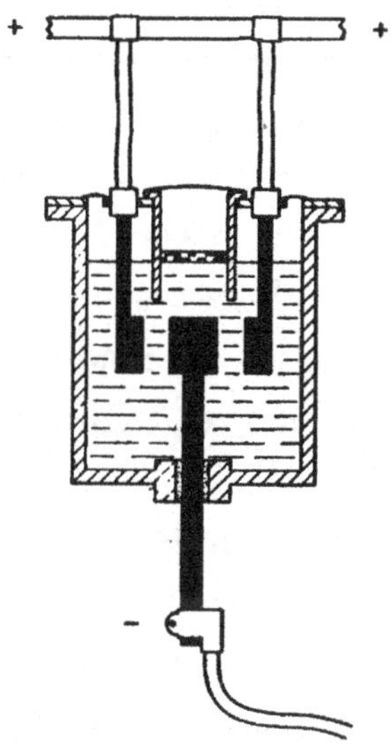

FIG. 73. Vessel for the electrolysis of caustic soda

203. Sodium hydroxide. We are already well acquainted with this substance from previous experiments and discussions. It is made at the present time almost exclusively by the electrolysis of sodium chloride. When this salt is dissolved in water the ions Na^+ and Cl^- are produced, and when an electric current is sent through the solution, sodium ion goes to the kathode and chloride ion to the anode. The chlorine leaves the ionic state and appears as chlorine gas, which is used for the manufacture of bleaching powder by passing it over wet lime. The sodium reacts at the kathode with the water of the solution, setting free hydrogen gas and forming sodium hydroxide.

The reactions are:

$$NaCl = Na^+ + Cl^-,$$
$$2\,Na^+ + 2\,Cl^- = 2\,Na + Cl_2,$$
$$2\,Na + 2\,H_2O = 2\,NaOH + H_2.$$

In order to prevent the sodium and chlorine from recombining, the vessel in which the electrolysis takes place is divided into two parts by a porous wall which permits the passage of the current, that is, of the ion carrying the current, but not the passage of the liquid or the gases. Since chlorine combines very energetically with almost all substances, the anode of this electrolytic cell must be made of platinum or graphite. The kathode is made of iron, which is not attacked very rapidly by the alkali.

A great many factories, paper mills for example, use in their processes both bleaching powder and caustic soda, and can therefore make immediate use of all the products of this electrolysis of sodium chloride except the hydrogen, and even this is sometimes used for heating purposes. In these factories the process is not carried to the production of solid sodium hydroxide. The solution formed in the kathode side of the cell is pumped directly into the tanks where caustic soda is needed. The chlorine produced at the anode is passed directly into limewater to form "bleach."

204. By-products. Where an arrangement of this kind is possible all the products can be made use of directly in the manufacturing process, and the cost of the chemicals is reduced to a very low figure indeed. This idea is called "making use of by-products," and it is the aim of every chemical industry to arrange for the use of as many as possible of the secondary products it produces.

205. The Le Blanc process. Sodium carbonate, or *soda* (Na_2CO_3), is made commercially by two very different processes. As we know, carbonic acid is weak and hydrochloric acid is very strong, so it is quite impossible to decompose sodium chloride by carbonic acid. A roundabout method is necessary, and many years ago the Le Blanc process was devised. It is still in use. The process depends on a rather complicated set of reactions.

First, sodium sulphate is made from sodium chloride by the action of strong sulphuric acid, according to the reaction

$$2\,NaCl + H_2SO_4 = 2\,HCl + Na_2SO_4,$$

with which we are already familiar.

The sodium sulphate is then heated with coal and the reaction is

$$Na_2SO_4 + 4\,C = Na_2S + 4\,CO.$$

Limestone is then added and more heat applied. The reaction is

$$Na_2S + CaCO_3 = CaS + Na_2CO_3.$$

The soda is then separated from the calcium sulphide by extraction with water and recrystallization, since the calcium sulphide which is formed under these conditions is very slightly soluble, while the soda is easily soluble.

In practice the process is shortened to two steps, the sodium sulphate being mixed with both coke and limestone. The last two reactions then go on together.

The principal by-product formed is hydrochloric acid, which results from the first step of the process. If it were not for the value of this by-product, the Le Blanc process would not be able to hold its own in point of cheapness with another process which is now being used for the manufacture of soda. This is the Solvay process, and both of these are used on an enormous scale in Germany and England. We shall study the reactions of the Solvay process in the chapter on the ammonium salts (sect. 225).

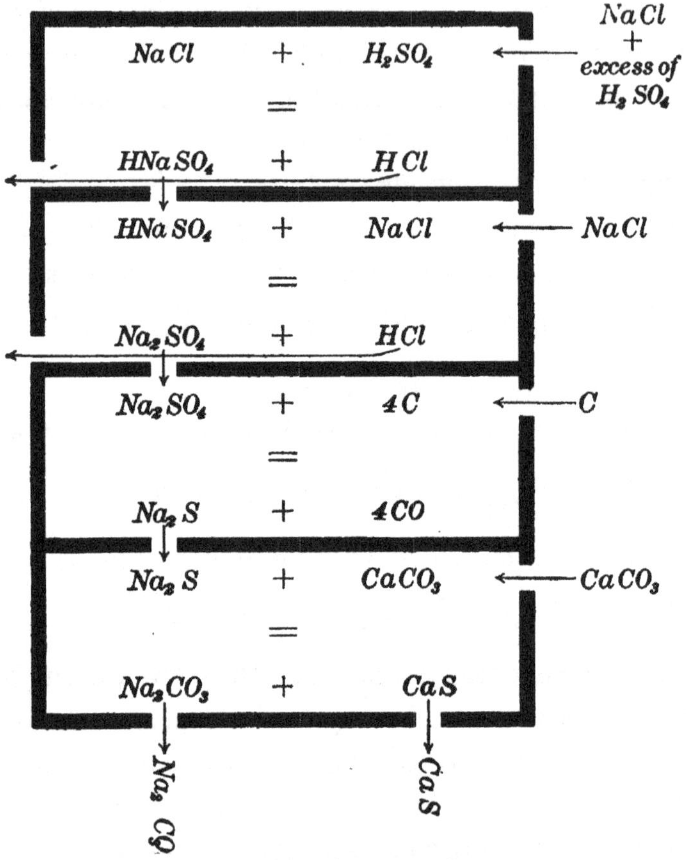

FIG. 74. Diagram of the Le Blanc process

206. Electrolytic soda. When carbon dioxide gas is passed into a solution of sodium hydroxide, sodium carbonate is formed, and this is the basis for a third method of making soda. The hydroxide is first made by electrolysis and then the carbonate is made from this. The process is usually carried out in such a way that acid sodium carbonate is formed instead of neutral sodium carbonate.

Experiment 147. Saturate a strong solution of neutral sodium carbonate (Na_2CO_3) with carbon dioxide.

Acid sodium carbonate (also called sodium bicarbonate) is formed according to the reaction

$$Na_2CO_3 + CO_2 + H_2O = 2\ NaHCO_3,$$

and this salt is very much less soluble than the neutral carbonate, so the solution does not need to be evaporated in order to cause the salt to crystallize out. This reaction can evidently be carried out just as well by passing carbon dioxide into caustic soda solution. In this case the neutral carbonate, which is very soluble, is first formed, and the bicarbonate is then produced from this by the action of more carbon dioxide.

207. Cooking soda. When the bicarbonate is heated or mixed with an acid or an acid salt, it gives out carbon dioxide, and this property gives it its value in cooking, so that it is very often called cooking soda. Its value in baking powders depends on the production of carbon dioxide when it reacts with an acid salt, the salt most often used being "cream of tartar." The *rising* of pastry and bread is caused by the escape of carbon dioxide in the heated mixture. Powders which are used to make effervescing drinks ("Seidlitz powders") contain sodium bicarbonate in one powder and tartaric acid or one of its acid salts in the other, and when these are mixed in water, carbon dioxide is liberated.

Crystalline sodium carbonate, $Na_2CO_3 + 10\ H_2O$, is very often called "alkali."

Experiment 148. Test solution of sodium carbonate with litmus paper. Try a solution of sodium bicarbonate in the same way.

Carbonic acid is so weak that its salts separate into acid and base in water solution, and since the hydrogen ion concentration is very small the base shows its usual reaction to litmus.

208. Hydrolysis. When an acid is very weak its ions are only present in very small concentration in solution. In the case of carbonic acid this means that the concentration of H^+ ion and the concentration of CO_3^{--} ion in a water solution of carbon dioxide is very small. If we add H^+ ions from without (by adding another stronger acid, for example), H_2CO_3 will be formed from its ions.

In the case of silicic acid (Exp. 145), when the acid was produced from its ions in this way it separated as a colloid from the solution. Carbonic acid has, however, no existence out of solution, and so the effect in this case would be to set free carbon dioxide as a gas.

$$2\,H^+ + CO_3^{--} = H_2CO_3,$$
$$H_2CO_3 = H_2O + CO_2.$$

Water itself contains H^+ ion and OH^- ion, but in very small concentration indeed. When we dissolve a salt of a very weak acid, like carbonic acid, in water, it forms ions in this way:

$$Na_2CO_3 = 2\,Na^+ + CO_3^{--}.$$

Whatever H^+ ion is present in the water itself combines with the CO_3^{--} ion to form nonionic carbonic acid (H_2CO_3), until the same concentration of H^+ ion and of CO_3^{--} ion is reached as would be formed by dissolving carbon dioxide in water to the same total concentration of H_2CO_3.

But these changes leave the OH^- ion of the water free. It cannot combine with Na^+ to form NaOH, for this substance is broken up into its ions quite perfectly in solution, so it remains in solution and shows its presence by giving all the tests for a base. This action of the ions of water on dissolved salts is called *hydrolysis*, and we shall find it again very frequently.

A solution of sodium bicarbonate ($NaHCO_3$) is exactly neutral to test paper, for here the hydrolytic action of the ions of the water (which gives OH^- ion) is just balanced by the small excess of hydrogen ion of the salt.

It will be seen from this that the words "neutral" and "acid" may mean different things, depending on the special property which we examine. What is called an acid salt is in this case neutral to litmus. It contains hydrogen, however, which is replaceable by a metal, and, judged by this property, which is a characteristic one for acids, sodium bicarbonate is an *acid* salt.

209. Sodium sulphate. This is another very important salt, which is made in great quantities in the first step of the Le Blanc process. It is especially interesting as an example of certain general properties.

The solubility of sodium sulphate in water increases with rise of temperature up to 32°. From that point on its solubility decreases with increase of temperature, so the solubility curve has the form shown in Figure 75. It will be noticed that instead of being a smooth curve, it rises to a point at 32°, and such a curve as this is an immediate indication that the two parts of the curve, on either side of 32°, express properties belonging to *two different substances.*

Experiment 149. Fill a small flask about a third full of sodium sulphate ($Na_2SO_4 + 10\ H_2O$), and warm it slowly on a water bath, stirring with a thermometer. Observe what takes place as the temperature rises, and the course of the temperature as shown by the thermometer.

At 32° Glauber's salt (sodium sulphate with 10 molecules of water of crystallization) breaks up into anhydrous (water free) sodium sulphate (Na_2SO_4), water, and a saturated solution of $Na_2SO_4 + 10\ H_2O$. It was the anhydrous salt which remained in the flask at temperatures above 32°. Glauber's

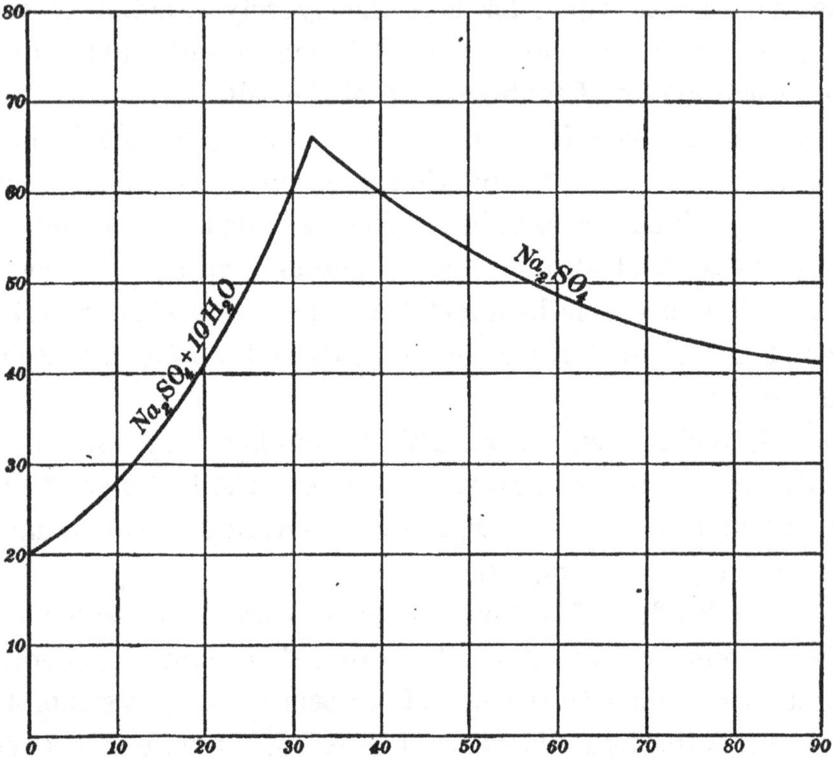

FIG. 75. Solubility curve of sodium sulphate

salt can only exist at temperatures up to 32°. This salt and the anhydrous sodium sulphate are two different substances, and so they have different solubilities in water. The first part of the curve shows the relation of temperature and solubility for Glauber's salt, while the part of the curve beyond 32° shows the same relation for the anhydrous sodium sulphate.

The experiments from which this curve was made were carried out by saturating water at various temperatures with sodium sulphate and then determining the amount of the substance in solution. There was always an excess of solid sodium sulphate left in the flask in which these solubility determinations were made, but this solid was $Na_2SO_4 + 10\ H_2O$ for all temperatures up to 32°, and Na_2SO_4 without water of crystallization at all temperatures above that point.

The same ions were of course present in the solution at all temperatures, Na^+ and SO_4^{--}, but when we determine the solubility of a substance we mean that equilibrium is reached between the solid we are shaking up with water in the flask and the ions of the same substance in solution. The equilibrium between Glauber's salt and the two ions Na^+ and SO_4^{--} is an entirely different one from that between anhydrous sodium sulphate and the same pair of ions.

Many substances show a solubility curve even more complex than this one, some of them having half a dozen points like the one at 32°. A point like this always indicates that the solubility measured on either side of it refers to two different substances, usually two substances differing in the amount of water of crystallization they contain.

210. Water of crystallization. Many salts which form good crystals take up water when they crystallize from solution, and this water forms a very definite part of the crystal and is written as a part of its chemical formula.

Experiment 150. Heat some copper sulphate crystals in a hard-glass test tube. Does copper sulphate crystallize with water of crystallization?

211. "Hypo." Another sodium salt which every one who takes photographs has occasion to use is sodium thiosulphate, or, as it is usually called, sodium hyposulphite, or *hypo*. It has the formula $Na_2S_2O_3 + 5\ H_2O$. It is made from sodium sulphite and sulphur. This salt has the property of easily dissolving silver chloride and silver bromide, and its use in photography depends on this fact.

Sodium thiosulphate melts easily to a clear liquid without the separation of any anhydrous salt. This is due simply to the fact that it is much more easily soluble than Glauber's salt, and needs less water to get it into solution than is present as water of crystallization in the crystals, provided, of course, the temperature is raised high enough to cause solution.

Experiment 151. Melt some sodium thiosulphate in a test tube, stoppered with cotton wool so that no dust can enter. After the salt is all melted let it cool down to room temperature. After tube and salt are cool rub a glass rod on a crystal of sodium thiosulphate and touch the supercooled salt in a tube with the rod.

The melting point of the substance is 56°, but it can exist as a liquid at very much lower temperatures if all particles of the solid are carefully excluded. The mouth of the tube must be closed with cotton wool to avoid the possibility of fine particles of the salt entering as dust, for as soon as the finest particle of the solid comes in contact with the supercooled liquid the latter is unstable and goes over into the more stable crystalline form.

212. Tests for sodium. Nearly all of the salts of sodium are easily soluble in water, so there is no easy way of separating it from other substances which are present as ions. Silver is easily removed from solution by bringing together silver ion and chloride ion; copper by bringing together copper ion and sulphide ion to form very slightly soluble copper sulphide; but we have no such reactions for sodium. There is, however, one test for sodium which is very sensitive indeed, and which distinguishes it absolutely from all other substances. That is the light which all sodium compounds as well as the element itself give out when they are heated to a high temperature.

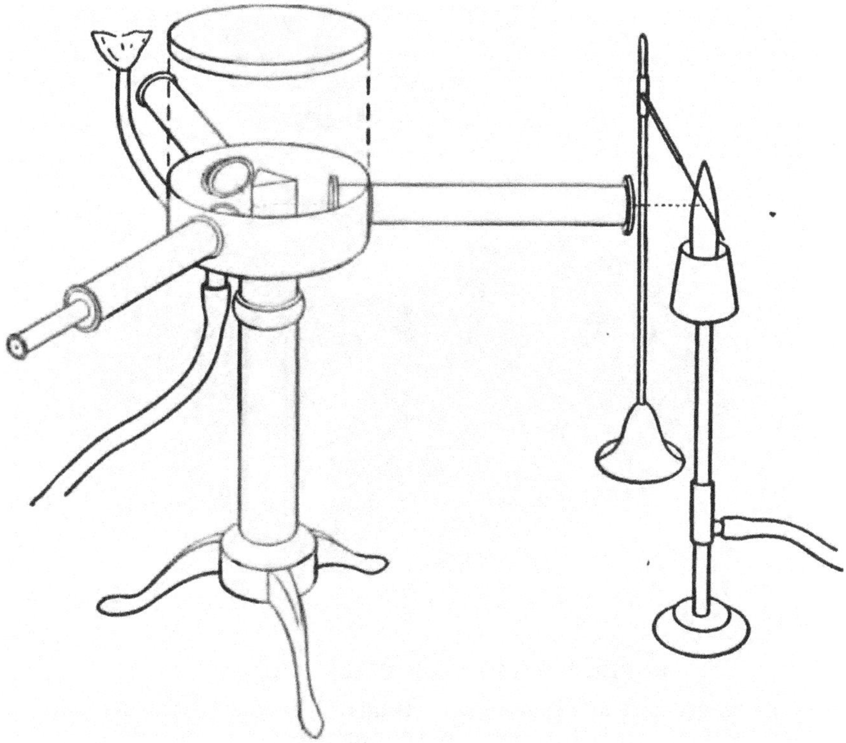

FIG. 76. Spectroscope arranged for testing a flame for sodium

Experiment 152. Hold a piece of platinum wire in the colorless flame of the burner until all color disappears from the flame. It will probably be necessary to use a new burner and a clean piece of wire to avoid impurities.

Touch the wire to a sodium salt and bring it into the flame again.

The yellow color is a test for sodium, and so delicate a test that all the flames in a laboratory where sodium salts have been used show traces of the yellow color.

ROBERT WILHELM BUNSEN (1811-1899)

One of the greatest of experimental chemists. Founded (with Kirchoff) the science of spectroscopy

213. Spectrum analysis. Much more exact than the method of merely observing the color of the flame is the test for sodium by means of the spectroscope. This instrument analyzes the light from the flame and shows a sharp, bright, yellow line when sodium is present, even though the flame is colored by other substances.

Many other substances show characteristic lines when examined in this way with the spectroscope, some of them

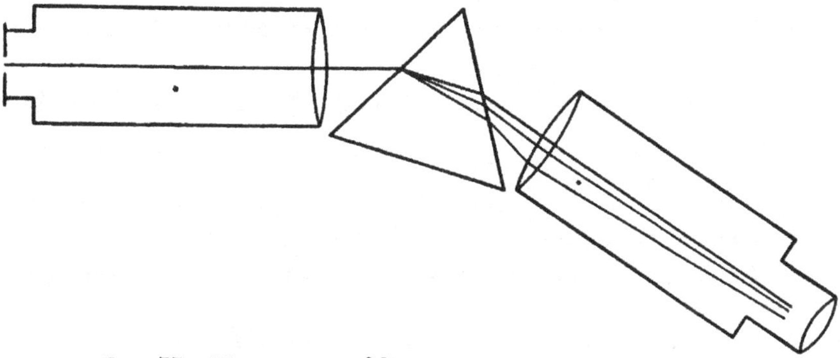

FIG. 77. The course of light through the spectroscope

only a single line, like sodium, and others a great many lines of different colors, but none of the lines from different elements coincide exactly, and so the presence of several elements in the same substance can be detected. Unknown substances can be analyzed by observing the lines they produce in the spectroscope and then trying known substances until one giving the same lines is found.

QUESTIONS

1. What are the general properties of metals?
2. Which of these properties are most important and characteristic?
3. What is our principal source of sodium?
4. How is common salt purified for table use?
5. How is metallic sodium made?

6. How is caustic soda made?

7. Explain why a plant for the electrolysis of common salt is an economy in a works where both caustic soda and bleaching powder are used.

8. What do these terms mean: soda, washing soda, baking soda?

9. What are the two principal processes by which sodium carbonate is made commercially?

10. What other process is also in use?

11. What by-products are produced by the Le Blanc process?

12. Describe the cycle of reactions which make up the Le Blanc process.

13. What are baking powders?

14. Why is crystalline sodium carbonate called "alkali"?

15. What is meant by hydrolysis?

16. How is it that a salt which is neutral to litmus paper may be an acid salt?

17. How is it that a salt which is alkaline in solution may be a neutral salt?

18. Explain the solubility curve of sodium sulphate.

19. Explain what takes place when Glauber's salt is slowly warmed on a water bath.

20. What are the properties of sodium thiosulphate?

21. Explain why sodium thiosulphate melts to a clear liquid and Glauber's salt does not.

22. What does the spectrum of sodium look like?

PROBLEMS

1. What weight of sodium hydroxide can be prepared from 100 gm. of sodium chloride?

2. What volume of hydrogen will be set free during this reaction?

3. How much sodiumcarbonate can be made from 100 gm. of sodium chloride?

4. If the Le Blanc process is used, what weight of hydrochloric acid will be produced during the treatment of 100 gm. of sodium chloride?

5. What volume of carbon monoxide will be set free during this process?

6. What volume of carbon dioxide must be passed into a solution containing 100 gm. of neutral sodium carbonate to convert it into sodium bicarbonate? What weight of sodium bicarbonate will be produced during this process?

CHAPTER XXIV

POTASSIUM

214. Properties. This element is very similar to sodium in all of its properties. It is a light, silver-white metal, almost as soft as wax. It oxidizes very rapidly in the air, its freshly cut surface becoming covered with a dull layer of oxide; and it acts on water so energetically that the hydrogen which is formed catches fire and burns with a reddish flame. The color of this flame is a test for potassium and its compounds, but it is easily obscured by the colors of other substances in the flame, and especially by the strong yellow light of sodium. The spectroscope gives a perfect means of overcoming this difficulty, and another way of avoiding it is to look at the flame through a piece of glass so colored that it absorbs the yellow sodium light but not the red light of potassium.

Experiment 153. Observe a flame in which both sodium and potassium salts are being volatilized, with and without the aid of a piece of blue cobalt glass. Could you distinguish these two elements from each other in a mixture?

215. Potassium hydroxide. This is formed when the metal acts on water, and it can be obtained in solid form by evaporating the solution. Solid caustic potash looks just like caustic soda, and it is an equally strong base when dissolved in water, so whenever hydroxide ion is needed for any reaction, either substance can replace the other with the same result.

Potassium salts occur in nature in vast amount in the feldspars and other silicates. Potassium chloride, the salt corresponding to sodium chloride, does not occur in as great quantities as the latter. Most potassium salts are made from the *sylvite* (potassium chloride) and *carnallite* (potassium magnesium chloride), which occur in great beds at Stassfurt in Germany.

216. Metallic potassium. The metal is made by electrolysis, but it is more difficult to manufacture than sodium and requires much more energy to get it out of combination, so its price is correspondingly high. The older way of making metallic potassium was like that for sodium, — heating the carbonate with carbon. The reaction is

$$K_2CO_3 + 2\,C = 2\,K + 3\,CO.$$

217. Potassium ion and sodium ion. Although sodium and potassium compounds can very frequently replace each other in chemical processes, this is only possible when the *anion* is the important thing for the reaction. The two metallic ions are entirely different in some of their reactions. Most plants need potassium for their life processes, and are wholly unable to make use of sodium in its place. This fact is evidently very closely connected with the fact that potassium salts are usually present in *soils*.

When potassium salts come in contact with the amorphous substances of the soil, potassium ion is held bound just as water is held by a plate of glass, while sodium ion is not so held (as mercury is not held by glass). The result is that the sodium salts which are set free during the decomposition of rocks are all washed out and soon find their way down to the sea. Potassium salts are kept back in the earth, and only a small amount of potassium is found in sea water. Plants can therefore always find potassium salts in soil, but cannot depend on sodium ion, and they have learned to make use of the ion which they are sure to find. Sea plants, and even those which live on salty ground or near the sea, use sodium in place of potassium ion.

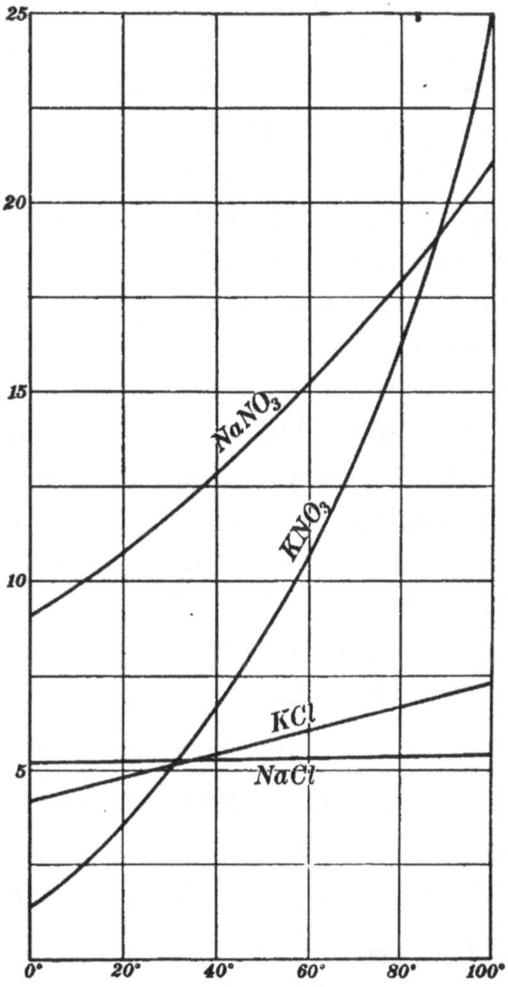

FIG. 78. Solubility curves

218. Saltpeter. Potassium nitrate is commonly called saltpeter. This salt has a very steep temperature—solubility curve; for at 0° 100 parts of water dissolve only 13 parts of KNO_3, while at 100° the same quantity of water will dissolve 247 parts. The curve shows these facts clearly (see Fig. 78).

The use of potassium in the manufacture of gunpowder depends on its large content of oxygen, and by the aid of this oxygen the sulphur and charcoal, which are the other two constituents of gunpowder, burn to form carbon dioxide and sulphur dioxide, even without the presence of air. The volume of gas produced by this combustion is very large compared with the volume of the gunpowder used to produce it, and, in addition to this, the gases are heated very hot by the large amount of heat set free, and the pressure is still further increased.

219. Explosion. A still more important fact, which is at the basis of all explosions, is that the reaction is very much accelerated by increase of temperature. The spark from the cap produces a high temperature at one point in a mass of gunpowder. Combustion is rapid at that point, and heat is produced which raises the temperature of the powder near by. This increase in temperature makes the reaction go faster still, and more and more heat is produced until the reaction goes so fast that it becomes an *explosion.*

Sodium nitrate would be cheaper than saltpeter for the manufacture of gunpowder, but it takes up water very easily (see sect. 165), and powder made from it must be kept carefully dry or it will spoil.

220. Making saltpeter. Much of the saltpeter used is made from sodium nitrate by the reaction

$$NaNO_3 + KCl = NaCl + KNO_3.$$

The ions in solution are the same on both sides of the equation,

$$Na^+ + NO_3^- + K^+ + Cl^- = Na^+ + NO_3^- + K^+ + Cl^-,$$

but potassic nitrate increases its solubility very rapidly with rise of temperature (see curve, Fig. 78), while sodium chloride is almost as soluble at low temperatures as at high (see curve). The solution containing these four ions is therefore evaporated until sodium chloride separates at the high temperature, and this is filtered off, the solution being kept hot during the filtration. The solution is then allowed to cool. Only a very little sodium chloride separates, but nearly all of the saltpeter will crystallize out at the lower temperature. In very many technical processes advantage is taken of the shape of the solubility curve of a substance as an aid to its separation and purification.

221. Bromides and iodides. Potassium bromide and potassium iodide are the usual forms in which bromide ion and iodide ion are used. Other soluble bromides and iodides could be used just as well, but most of them have inconvenient properties. Calcium iodide cannot be used, although calcium compounds are generally cheaper than the corresponding potassium ones, because calcium iodide is very deliquescent. The principal cost is in the iodine, and the difference in cost of the kation disappears in comparison.

Iodine is only very slightly soluble in water, but it is very soluble in potassium iodide solution. It is a general rule that when a substance which is only slightly soluble in water is more soluble in a solution of a salt, it means that a *new ion* is formed from one of the ions of the salt and the substance which goes into solution. In this case iodine combines with the iodide ion (I^-) to form a new ion of symbol I_3^- (tri-iodide ion), and this has the property of breaking up again very easily to form iodine and iodide ion (I^-). A solution of iodine in potassium iodide can therefore be used for almost all purposes where we would ordinarily use pure iodine, with the further advantage that it can be had in any desired concentration.

222. Reagent for potassium ion. Like the sodium salts, almost all the salts of potassium are easily soluble in water. There is, however, one which can be used to separate potassium from sodium, since it is not very soluble in water and still less so in alcohol. It is the potassium salt of a complex acid which contains platinum.

Experiment 154. Add hydrochlorplatinic acid or sodium platinic chloride to a little of a solution of a potassium salt. Filter off the precipitate and add alcohol to the filtrate.

223. Potash. Potassium carbonate was at one time a more important substance than it is now. A solution of this salt in water is quite alkaline (Why ?), and it was formerly made by treating wood ashes with water and evaporating the solution. The crude salt was called *potash*. It is still used in chemical processes where potassium is necessary and the volatile nature of carbon dioxide an advantage. Since the development of the great soda processes it has been replaced for most purposes by sodium carbonate.

QUESTIONS

1. What are the properties of metallic potassium ?

2. What is the color of the potassium flame ?

3. How could you show the presence of potassium in the flame when sodium is also present ?

4. Why is it that sodium and potassium salts are usually interchangeable in the laboratory ?

5. Why do sodium salts not make good fertilizers ?

6. What is gunpowder ?

7. Describe the course of an explosion, showing how increase in reaction velocity and increase in temperature are dependent on each other.

8. How can potassium nitrate be made from sodium nitrate ? Explain with the aid of the solubility curves of these two substances.

9. Why is iodine more soluble in potassium iodide solution than in water ?

10. What precipitation test can be used for potassium ion ?

PROBLEM

1. What weight of potassium nitrate can be made from 100 gm. of sodium nitrate ?

CHAPTER XXV

AMMONIUM

224. Salts of ammonium. In Chapter XX we learned that a solution of ammonia gas in water shows all the properties of a base, which cannot be prepared in the free state, but which forms a set of stable salts containing the ion NH_4. The salts of ammonium are colorless and very similar to those of potassium in their solubilities and even in their crystalline forms.

225. The Solvay process. In section 205 it was mentioned that one of the important processes for making soda, called the Solvay process, is based on ammonium compounds. The chemistry of this process is so interesting that we shall take it up quite fully.

First of all, a solution of common salt is mixed with ammonium carbonate, and carbon dioxide is passed into the solution. The following reaction takes place:

$$2\,NaCl + (NH_4)_2CO_3 + CO_2 + H_2O = 2\,NaHCO_3 + 2\,NH_4Cl.$$

There are a number of things to be considered in this reaction. The solution contains the ions

$$Na^+,\ Cl^-,\ NH_4^+,\ \text{and}\ HCO_3^-,$$

and the first salt to crystallize out of a solution containing these four ions will be the least soluble one which can be formed by combining them. The least soluble salt which can

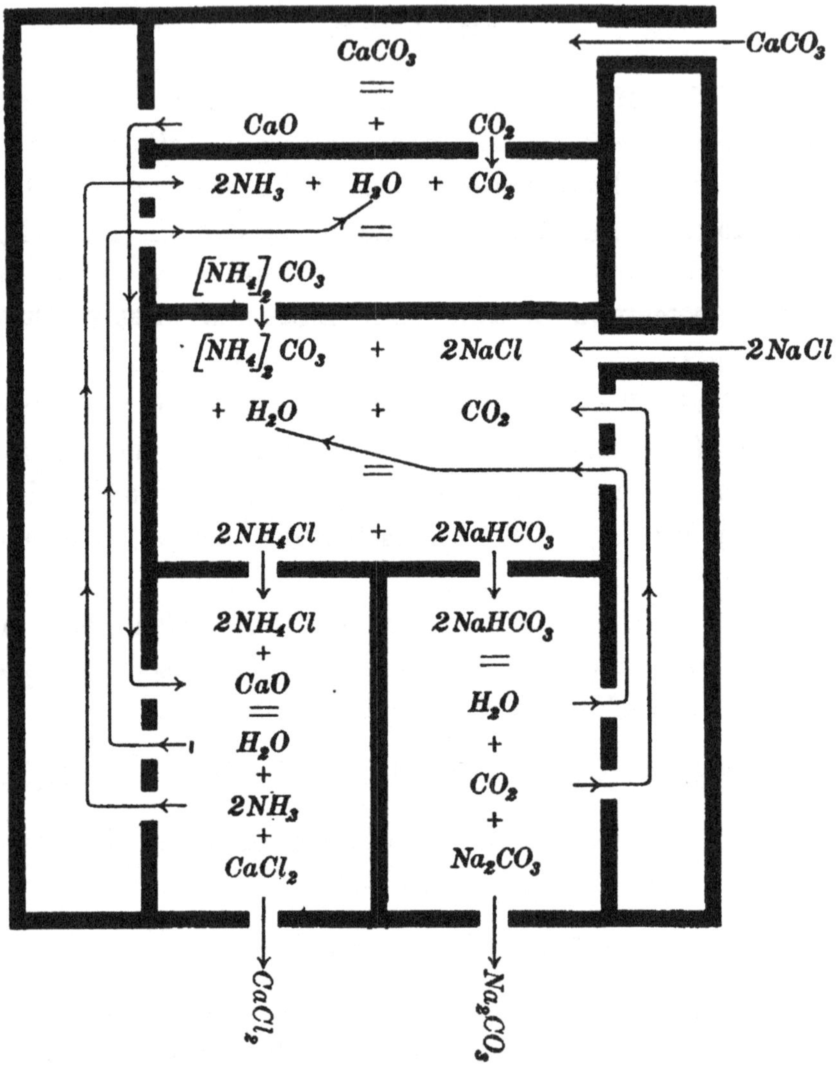

FIG. 79. Diagram of the Solvay process

form from any combination of these ions is sodium bicarbonate ($NaHCO_3$), and so this is the substance which separates out in solid form.

In order to make the process a cheap one all the ammonia must be recovered, and we must, beside, change most of the carbonate into neutral sodium carbonate, as the demand for this is much greater and it is the real object of the process. The ammonia can be recovered by adding lime to the ammonium chloride which remains in solution. Ammonia gas escapes, and this is condensed and used over again, while calcium chloride remains behind. The reaction is

$$2\,NH_4Cl + CaO = 2\,NH_3 + CaCl_2 + H_2O.$$

The sodium bicarbonate is changed into neutral carbonate by simply heating it, giving the reaction

$$2\,NaHCO_3 = Na_2CO_3 + CO_2 + H_2O.$$

The carbon dioxide which escapes is also recovered and used again, but only half enough is recovered to continue the process, since the other half is sent to market as sodium carbonate. The other half of the carbon dioxide comes from limestone, which is heated to produce it, and the lime which is left behind is used in the recovery of the ammonia. This ammonia is recombined with carbon dioxide to form ammonium carbonate, and the process can then begin again on a new supply of sodium chloride.

The final reaction which takes place is then

$$2\,NaCl + CaCO_3 = Na_2CO_3 + CaCl_2,$$

and this means that if we arrange by careful manufacturing methods to recover all the ammonia, we need to take into the factory only common salt (NaCl) and limestone ($CaCO_3$), and there will come out at the other door soda (Na_2CO_3) and calcium chloride ($CaCl_2$).

The ammonium carbonate, which is a necessary part of the process, never goes outside the doors of the factory, but simply around in a circle. The principal property which makes this process possible is the volatility of ammonia.

226. By-products. If the soda process depended on the market price of sodium carbonate alone, the Solvay process would soon drive all the more complicated ones out of business. But the by-products of the Le Blanc process, the most important of which are hydrochloric acid and sodium thiosulphate, are valuable enough to enable the process to hold its own in some parts of the world.

It is easy to see how these great manufacturing industries have lowered the price of hydrochloric acid, sodium thiosulphate, and calcium chloride, since these by-products are formed in immense quantities. It is upon relations of this kind that the market price of any chemical depends.

QUESTIONS

1. Tell what you can of the Solvay process for making soda.

2. Explain how the market price of any chemical may depend on great manufacturing industries like the Le Blanc and Solvay processes.

PROBLEMS

1. In the Solvay process how many parts of lime must be supplied for each 100 parts of sodium chloride treated?

2. What weight of calcium chloride is produced in this process for each 100 gm. of sodium chloride treated?

CHAPTER XXVI

CALCIUM

Barium, Strontium, Magnesium

227. The metal calcium. Metallic calcium has only been prepared in large quantities during the past few years, but now electrolytic processes have been found which make its manufacture fairly cheap. It is a rather hard metal of the color of steel, which combines with oxygen at ordinary temperatures, but not with the avidity of potassium or sodium. So far it has found only limited technical application, but its cheapness will probably bring uses for it before long.

Experiment 155. Try the action of metallic calcium on water. Is a base formed?

228. Calcium carbonate minerals. Calcium occurs in huge masses in nature in the form of limestone. It also occurs as marbles of various colors, as calcite, chalk, and Iceland spar. Calcium compounds are, after those of silicon, aluminium, and iron, the most common constituents of rocks (see page 282). All the minerals mentioned are calcium carbonate.

Experiment 156. Add hydrochloric acid to limestone, marble, chalk, calcite. What gas is formed? Write the reaction.

229. Lime. When these minerals are heated carbon dioxide is driven off according to the reaction

$$CaCO_3 = CaO + CO_2.$$

The substance CaO is calcium oxide, usually called lime, and it is made by heating limestone until the carbon dioxide is all driven out.

Experiment 157. Heat some water to boiling. Remove the flame and add some small bits of lime. Test the solution with litmus.

This process is called *slaking* lime, and the chemical reaction which takes place is

$$CaO + H_2O = Ca(OH)_2.$$

The substance produced is calcium hydroxide (slaked lime), and calcium oxide is its anhydride, since it combines with water to form it. So much heat is produced during the combination that the water boils.

230. Mortar. Lime is used in large quantities in the making of *mortar*. This is made by first slaking lime and then mixing it with sand; the mixture is then spread between the stones or bricks which it is to bind together, or over the wall which it is to cover. The *setting* of mortar depends on the carbon dioxide of the air, which is taken up by the mortar with the formation of calcium carbonate. This crystallizes slowly into a substance as hard as rock. When the wet mortar is laid on, it enters the little cavities of the stones and bricks, and when it finally hardens, the whole mass is firmly bound together. If mortar is made without sand, it contracts too much and leaves cracks and holes; but if a large part of the space to be occupied by mortar is filled with grains of sand, these are bound together and assist in the setting of the wall.

The formula for the process is

$$Ca(OH)_2 + CO_2 = CaCO_3 + H_2O.$$

Water is set free during the process, and this explains why the walls of houses which have been newly plastered and which have apparently become quite dry, *sweat* when the family first moves into the house. The carbon dioxide exhaled by the people in the house causes this reaction to go on quite rapidly, and water is formed so fast that it appears in drops. To avoid this, houses must be aired for a long time after the walls are finished, or carbon dioxide must be produced by burning an open fire in the house, so that the reaction of the setting of the mortar may be quite complete before the house is occupied.

Experiment 158. Test a piece of old mortar from a wall for the presence of calcium carbonate.

231. Air-slaking. When lime is exposed to the air it takes up carbon dioxide and changes back to calcium carbonate again. This process is called air-slaking and it suggests an interesting question. Quicklime is produced from calcium carbonate by heating it until all the carbon dioxide is driven off. Calcium carbonate is formed from quicklime and carbon dioxide. The reaction which goes on at a high temperature goes on in exactly the reverse direction at a low temperature. What would take place at an intermediate point?

232. Equilibrium. The relation between calcium carbonate and carbon dioxide is somewhat like that between liquid water and water vapor. Water has at every temperature a definite *water-vapor pressure*, and calcium carbonate has at each temperature a definite *carbon dioxide pressure*, which increases with rise of temperature. If calcium carbonate is heated in a closed vessel, the carbon dioxide pressure increases until at about 815° it reaches atmospheric pressure. This would be a high enough temperature for the production of quicklime from limestone, but it is usually heated hotter than this, so that the process may go on rapidly.

Ordinary air contains only about one part of carbon dioxide in two thousand parts of air, so the carbon dioxide pressure of the air is only about one two-thousandth of an atmosphere. But even this small pressure is greater than the carbon dioxide pressure of calcium carbonate at room temperature, and so instead of giving off carbon dioxide, a mixture of lime and calcium carbonate takes up carbon dioxide from the air to form more calcium carbonate.

233. Reversible reactions. We have had other reactions like this. Sodium bicarbonate is made by passing carbon dioxide over moist sodium carbonate at room temperature. When the bicarbonate is heated it breaks up into neutral carbonate, carbon dioxide, and water. There is, then, for each temperature a definite equilibrium between these four substances, —neutral sodium carbonate, sodium bicarbonate, water, and carbon dioxide. If we increase the temperature slightly, carbon dioxide and water vapor escape until the pressure is just balanced at the new temperature. If we lower the temperature, carbon dioxide and water combine with the neutral carbonate to form bicarbonate until everything is balanced again.

These are examples of what is called chemical equilibrium, and almost any chemical reaction which has come to a stop at any temperature, or, as we can express it, has reached equilibrium, can be set in motion again in one direction or the other by a change in the temperature.

234. Gypsum and plaster of Paris. Another important calcium compound which occurs in nature is the mineral gypsum, which has the formula $CaSO_4 + 2\,H_2O$. When this substance is heated it loses its water of crystallization and falls to a powder, losing its crystalline structure completely. This white powder is what is known as "plaster of Paris," and when it is mixed with water this anhydrous calcium sulphate takes up its two molecules of water of crystallization again and changes back into crystalline form, hardening into a coherent mass. This property is made use of in making plaster casts. A mold is made of clay, which is allowed to set hard enough to hold its shape, and is then filled with a liquid mixture of plaster and water. This fills all the corners of the mold and in a few minutes *sets* or hardens, the anhydrous sulphate taking up the water with which it is mixed, and crystallizing to a hard mass. The mold is then removed and a cast hard enough to stand ordinary usage is ready.

235. Calcium ion. Calcium ion is colorless, as we have seen from its solutions, and it differs from all the other metallic ions we have taken up in being *bivalent*, calcium hydroxide having the formula $Ca(OH)_2$, the sulphate $CaSO_4$, the chloride $CaCl_2$, etc.

Calcium phosphate is another very valuable substance. Phosphoric acid has the formula H_3PO_4. It contains three replaceable hydrogens and can therefore form three different salts with calcium. The neutral calcium phosphate is $Ca_3(PO_4)_2$, and this is the calcium phosphate which occurs in nature and which is one of the necessary constituents of the bones of animals. The other phosphates of calcium are $CaHPO_4$ and $Ca(H_2PO_4)_2$. This last substance is a constituent of "superphosphates," which is used in fertilizers. It is formed from the naturally occurring phosphate by the action of sulphuric acid:

$$Ca_3(PO_4)_2 + H_2SO_4 = Ca(H_2PO_4)_2 + CaSO_4.$$

Monocalcium phosphate, as it is called, is easily soluble in water, and can therefore be easily taken up by the roots of plants, while the neutral phosphate is only very slightly soluble, and is therefore not directly useful in fertilizers.

236. Calcium chloride. This substance has been mentioned as a by-product in the Solvay process for making soda, and we have used it in several experiments as a drying agent. It takes up water very readily, and is used for all sorts of purposes where dry air or any dry gas is desired.

The drying power of a substance depends on the vapor pressure of the substance formed when water combines with the drying agent. Calcium chloride does not dry air which is passed through it as completely as strong sulphuric acid does, and air which has been passed through this strong acid can be dried still more completely by passing it through dry phosphorus pentoxide. But because calcium chloride is a very cheap substance indeed, and also because it does not attack things as strong sulphuric and phosphoric acids do, it is very generally used as a drying agent where the utmost dryness is not necessary.

237. How glass is made. Calcium compounds are used extensively in the manufacture of glass. Ordinary glass is made by heating together soda, sand, and limestone. Carbon dioxide is driven off and the silicates of sodium and calcium melt together at the high temperature of the glass furnace into a transparent liquid which cools to glass. This reaction takes place in spite of the fact that carbonic acid is a stronger acid in water solution than silicic acid, because the carbon dioxide escapes from the molten mass. Since this gas is escaping during the reaction, it is necessary to keep the mass fluid until all the bubbles of carbon dioxide can rise and escape. If a very clear, fine glass is desired, the liquid mass must be kept hot for a long time.

238. Calcium carbide. This is a remarkable calcium compound, which is made by heating lime and coal to a very high temperature in an electric furnace. The reaction is

$$CaO + 3\ C = CaC_2 + CO.$$

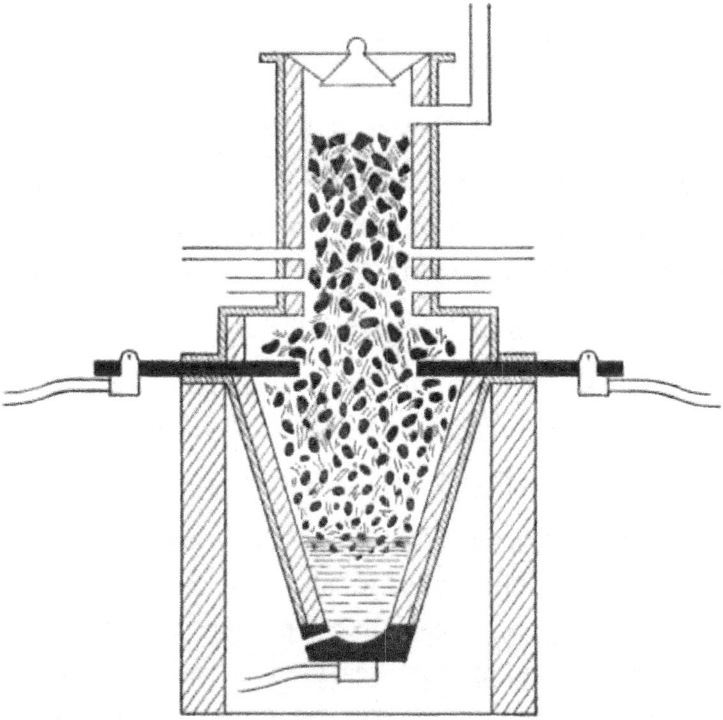

FIG. 80. Electric furnace for making calcium carbide

When calcium carbide comes in contact with water the reaction

$$CaC_2 + 2\ H_2O = Ca(OH)_2 + C_2H_2$$

takes place, and acetylene (C_2H_2) is formed. This is a gas which burns in air with the formation of carbon dioxide and water, giving at the same time a very bright light.

239. Thermochemistry. When carbon burns with oxygen the reaction is

$$2\,C + 2\,O_2 = 2\,CO_2 + 812\text{ KJ}.$$

812 measures the quantity of heat in thousands of joules (kilojoules) which is given out when 2 combining weights of carbon (2 × 12 gm.) combine with 4 combining weights of oxygen (4 × 16 gm.) to form 2 combining weights of carbon dioxide (2 × 44 gm.).

When hydrogen burns with oxygen to form water the reaction is

$$2\,H + O = H_2O + 286\text{ KJ}.$$

If we burn two combining weights of carbon and two of hydrogen separately, the quantity of heat set free is the sum of the two heats of reaction:

$$812 + 286 = 1098\text{ KJ}.$$

But if instead of burning the two elements separately we burn acetylene, the reaction is

$$C_2H_2 + 5\,O = 2\,CO_2 + H_2O + 1321\text{ KJ},$$

and 223 KJ more heat is set free. The quantities of carbon and hydrogen are the same in the two cases, but the compound furnishes more heat than its constituents.

These figures explain why acetylene gives out such a bright light when it burns. The extra heat makes the temperature of combustion higher, and the carbon particles which give out the light are heated hotter than in other flames.

The 223 KJ of extra heat must come from somewhere, and it does in fact require the expenditure of a great deal of energy to make acetylene, and still more to make the calcium carbide which we use to generate it. When we burn acetylene we get back not only all the heat of combustion of carbon and hydrogen but also a part at least of the energy which was used up in causing calcium and carbon to combine to calcium carbide.

240. Strontium and barium. Two other elements are very much like calcium in their chemical and physical properties. They are called strontium and barium.

Barium. We have often made use of barium ion (Ba^{++}) as a test for sulphate ion, because barium sulphate is a very slightly soluble substance. Ba^{++} ion is colorless and rather poisonous. Metallic barium has no technical application, and it is therefore only made in very small quantities for scientific purposes.

The principal barium compounds which occur in nature are the sulphate, which is called *barite* or *heavy spar,* and the carbonate, which is called *witherite.* All the compounds of barium are made from these. Barium sulphate is very slightly soluble, not only in water but also in strong acids. It is used as a paint under the name of "permanent white." The nitrate of barium is used in fireworks, to give a green color to flames. All barium salts give this green color and show a corresponding series of lines in the spectroscope, which give a very delicate test for the element.

Strontium salts give a red color to the flame, and they are used in making red fire. The metal itself is very much like calcium and barium. Sr^{++} ion is colorless. Strontium occurs in nature principally as sulphate and carbonate.

241. Magnesium. Magnesium is similar in many ways to these three metals. We already know a good deal about the magnesium salts, for we have used the metal as a reagent for hydrogen ion, and magnesium salts have been produced by the reaction. The metal is used to produce a very bright light, for its heat of combustion is large and the white oxide which is formed is heated very hot during the combustion and gives out a white, dazzling light.

All magnesium salts have a bitter taste, which is therefore a characteristic of magnesium ion (Mg^{++}), just as the sour taste of acids is a property of hydrogen ion. Compounds of magnesium occur in large masses on the earth. The double compound $CaCO_3$, $MgCO_3$ is called *dolomite*, and a great mountain range in Europe is composed almost entirely of this mineral.

The following table shows the properties of some of the usual compounds of the four metals studied in this chapter.

	MAGNESIUM	CALCIUM	STRONTIUM	BARIUM
Chloride . .	Soluble	Soluble	Soluble	Soluble
Sulphate . .	Soluble	Less soluble	Very slightly soluble	Insoluble
Hydroxide .	Insoluble	Very slightly soluble	Less soluble	Soluble
Metal	Rather easy to reduce	More difficult to reduce	More difficult to reduce	Very difficult to reduce

QUESTIONS

1. What are the properties of metallic calcium?
2. What are the properties of calcium hydroxide?
3. What common minerals are composed of calcium carbonate?
4. What is the reaction when calcium carbonate is heated?
5. What is slaked lime?
6. Why are bubbles and steam produced when quicklime is being slaked with water?
7. What is mortar? To what reaction is the setting of mortar due?
8. Why must houses be aired for some time after the walls are plastered?

9. Explain the relations existing between calcium carbonate, quicklime, and carbon dioxide at various temperatures.

10. What is gypsum?

11. What is plaster of Paris?

12. What process takes place when plaster sets?

13. Which of the calcium phosphates occurs in nature? Why does it not make a good fertilizer? How is it changed into a better form for fertilizing purposes?

14. What is the formula for the soluble calcium phosphate?

15. What determines the drying power of a substance?

16. How is glass made?

17. Why must the molten mass of glass be kept hot for a long time if a clear, fine glass is desired?

18. How is calcium carbide made?

19. What is the reaction between calcium carbide and water?

20. What are the properties of acetylene?

21. Explain why it is that we can get more heat by burning acetylene than by burning the same weight of carbon and hydrogen separately.

22. Tell what you can about strontium and its compounds.

23. Tell what you can about barium and its compounds.

24. Why do all magnesium compounds show the same bitter taste?

25. Tell what you can about metallic magnesium and its compounds.

PROBLEMS

1. What weight of lime can be produced from 1000 kgm. of calcium carbonate? What volume of carbon dioxide gas will be set free during this reaction?

2. What weight of water is freed during the setting of a kilogram of calcium hydroxide in mortar?

3. What weight of plaster of Paris can be made from a kilogram of pure gypsum?

4. What weight of sulphuric acid must be used in the treatment of a kilogram of calcium phosphate in the manufacture of phosphate for fertilizer?

CHAPTER XXVII

ALUMINIUM

242. Trivalent metals. The alkali metals are univalent; that means that sodium and potassium replace one combining weight of hydrogen in an acid to form salts. Calcium, barium, and strontium are bivalent, and replace two hydrogen atoms. The metals of the next group are *trivalent* and replace three hydrogen atoms. The only member of this group which is at all well known or important is aluminium; the other members are very rare metals, and we shall not study about them at all in this course.

243. The metal aluminium. Metallic aluminium has been known for about seventy years, but it is not many years since it became cheap enough to be used for any technical purpose. The improvement of dynamo machinery and the resulting cheapness of production of electric current have made possible the commercial manufacture of many substances which require great expenditure of energy in their preparation. This energy can often be provided in the form of heat from burning coal, but much is lost in this case because there is no way of avoiding radiation from hot bodies.

By using an electric current it is possible to provide great quantities of energy condensed into a small space, and so to produce a very high temperature indeed without having coal or any other fuel in contact with the reacting substance. This is not the only advantage of the current. It enables us to produce at the electrodes two very different chemical processes, — reduction at the kathode, oxidation at the anode. The production of a very high temperature and the special

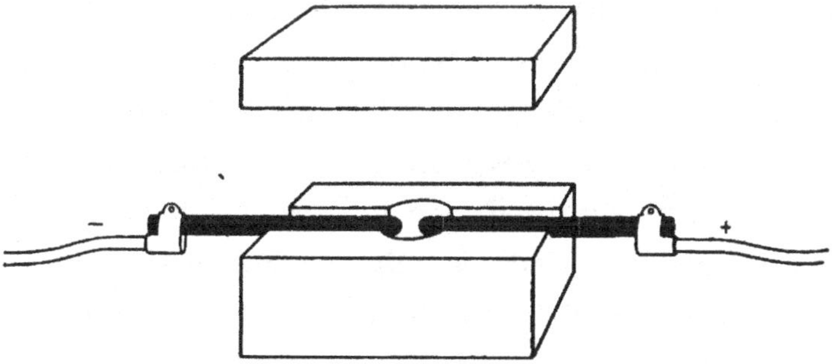

FIG. 81. Arc furnace

reduction and oxidation at the electrodes are two separate characteristics of electrical energy, which form the basis for its commercial applications.

244. Electric furnaces. To produce the high temperature we use *electric furnaces*, of which there are two kinds, — arc furnaces and resistance furnaces. If the arc of a street lamp is inclosed in a hollow block of some material which does not easily melt, we have an example of an arc furnace. Here all the energy of the current is transformed

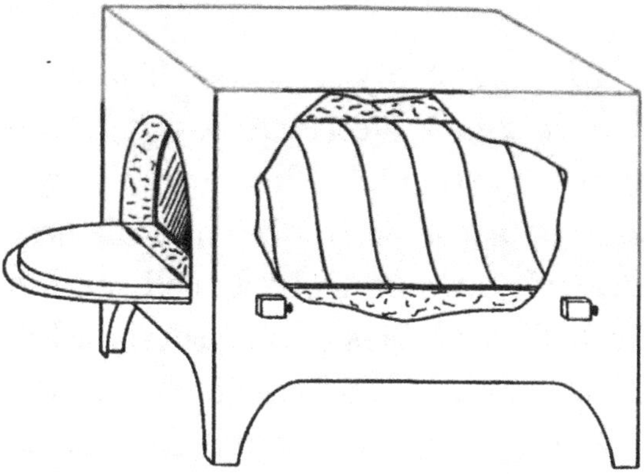

FIG. 82. Resistance furnace

into heat between the carbon terminals, and the temperature produced is a very high one, probably as high as 3000°. The disadvantages of this form are the extreme concentration of the heat and the fact that the products of combustion of carbon in air — carbon dioxide and other substances (cyanogen) — mix with the reacting materials. The other form of furnace is made by passing the current through a resisting material, which is heated to a white heat by the current. This resistance material may be placed in any way around the body of the furnace, either by winding it in the form of wire about a cylinder, or by using resisting masses which can be built up into any desired form with a hollow space in the center. The interior of furnaces of this type are heated without any products of combustion being added to the substance being treated.

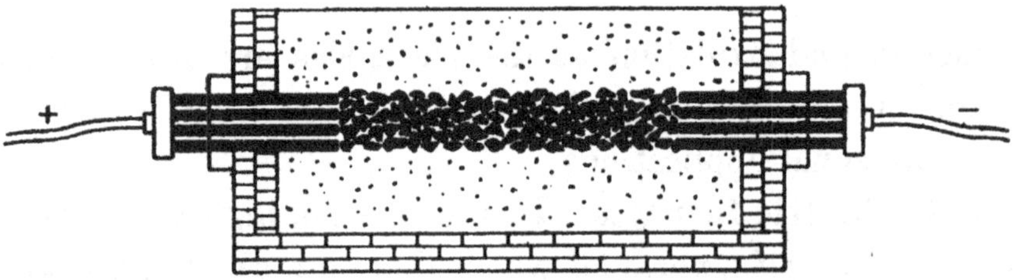

FIG. 83. Large commercial type of resistance furnace

245. Electrolytic cells. The other and wholly different way in which electrical energy is used provides for its direct transformation into chemical energy without the losses which always accompany the use of heat. The current is sent through an electrolytic cell and a kation moves to the kathode and is set free there, while at the same time an anion moves to the anode and is set free there. Such an arrangement we have used several times; in the decomposition of water, for example (Exp. 74), and in our experiments in electrolysis (Exps. 77, 79, 80). A similar arrangement is used for the production of metallic aluminium on a large scale. Here a water solution cannot be used and a fused compound of aluminium is substituted.

246. The manufacture of aluminium. Figure 84 shows the arrangement of the apparatus. The bottom of the cell is first covered with an easily fusible compound of aluminium called *cryolite*, which is a fluoride of sodium and aluminium. The anodes are then lowered so that they are close to the bottom of the cell, which is made the kathode. The cell is then filled with cryolite and the current turned on, melting the cryolite. As soon as the process is started, dry aluminium oxide (Al_2O_3)

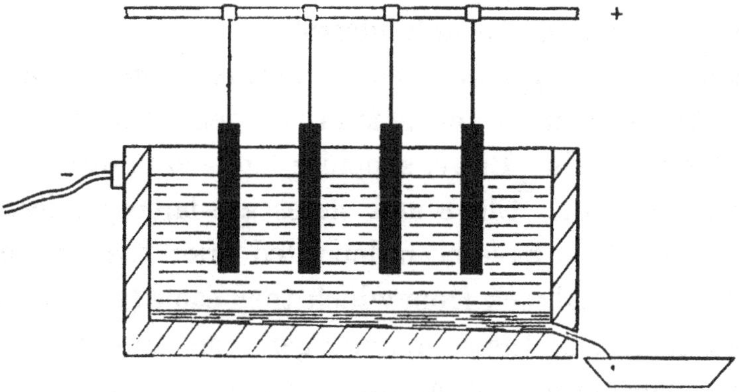

FIG. 84. Electrolytic manufacture of aluminium

is added, and this dissolves in the molten cryolite and is decomposed by the current into metallic aluminium, which collects in the bottom of the cell, and oxygen, which unites with the carbon of the anodes, forming carbon monoxide, which escapes and burns at the top of the cell. The aluminium oxide is added fast enough to supply the aluminium and oxygen as they are separated, and the cryolite acts only as a solvent. Improvements in the production of electrical energy and in its application to chemical manufacture have reduced the price of aluminium from twelve dollars a pound in 1887 to about twenty-five cents at the present time.

247. Properties. This metal never occurs in nature because it combines so very readily with oxygen. The only reason why vessels made of aluminium last for any length of time is because they become covered with a resisting layer of oxide, which is so hard and tough that air cannot get through it, and the rest of the metal is thus protected from oxidation.

Experiment 159. Examine the physical properties of aluminium. Rub some mercury on a clean surface of the metal.

Where the mercury prevents the formation of a hard film the oxidation can go on unhindered.

Aluminium compounds occur in nature in great quantities as silicates and oxides, and it is in this sense the most abundant of all metals. Ruby, sapphire, and corundum are oxides of aluminium. Clays and slates are largely silicates of aluminium, which have been deposited after the decomposition of the original rock from which they came (see sect. 196). A large number of the commoner gems are aluminium compounds. Turquoise is aluminium phosphate, and garnet contains aluminium.

Aluminium is attacked by acids and dissolves with the formation of salts. The reaction in the case of hydrochloric acid is

$$Al + HCl = AlCl_3 + 3\ H,$$

corresponding to the trivalent character of the metal.

Aluminium hydroxide is a very weak base indeed. Solutions of aluminium salts in water are split up in part into acid and base, as in the case of salts of carbonic acid and silicic acid. In those cases, however, it is the acid which is weak, and solutions of carbonates show the reaction for a base and turn red litmus paper blue. In this case it is the base which is weak, and the solutions of its salts are acid, turning blue litmus paper red.

Experiment 160. Test a solution of an aluminium salt with litmus paper.

248. Hydrolysis. As we learned in section 208, water contains in very small concentration both H^+ ion and OH^- ion. It can act, therefore, either as an acid or a base, but a very weak one in either case. But water can only show the properties of one of its ions when a substance is dissolved in it which removes the other from the solution. When aluminium chloride is dissolved in water the Al^{+++} ion which forms combines with the OH^- ion of the water to form $Al(OH)_3$. This is such a very weak base that there is less OH^- ion left in the solution than there was in the original pure water, and the H^+ ion which is left shows its acid nature by reaction with litmus. This is another example of *hydrolysis.*

If aluminium hydroxide is formed by adding caustic potash (or soda) to a solution of aluminium chloride (or any other aluminium salt), a rather strange thing happens.

Experiment 161. Add a little caustic potash solution to a solution of aluminium chloride. What are the properties of aluminium hydroxide? Keep on adding more and more caustic potash solution. What happens?

Aluminium hydroxide can act not only as a base, forming salts with strong acids like hydrochloric acid, but also as an acid, forming salts with strong bases like potassium hydroxide. Metallic aluminium reacts with caustic potash, setting hydrogen free.

Experiment 162. Cover some pieces of metallic aluminium with strong caustic potash solution and heat. What gas is produced?

The salt which is formed by this reaction is called potassium aluminate. It is a very general fact that when the basic properties of a hydroxide are very weak, weak acid properties also appear, and, as in the case of water, which shows acid or basic properties depending on which of its ions is removed, so these hydroxides which are weak bases show acid properties, combining with metals which form strong bases, to form salts.

249. Rocks. Many rocks are compounds of aluminium silicate with other silicates. The silicates of the other metals are decomposed by the carbon dioxide (and moisture) of the air with the formation of carbonates, which are usually soluble in carbonic acid solution, and are therefore washed out by water containing carbon dioxide.

Aluminium silicate is left behind, for carbonic acid is so weak an acid and aluminium hydroxide is so weak a base that no compound of the two is formed. No carbonate of aluminium has ever been prepared. The washing out of the other metals in the form of carbonates leaves the aluminium silicate behind in the form of a powder, which is carried away by water and deposited as clay or loam. These usually contain, beside aluminium silicate, quartz and iron oxide, and the latter colors them red or brown.

The purest form of aluminium silicate is *kaolin*, which is white, and this is used in the manufacture of fine porcelain. Other cheaper kinds of pottery are made of the more impure clays, which are baked until the water is all driven off.

The sulphate of aluminium forms crystals with 24 molecules of crystal water.

Experiment 163. Make saturated solutions of potassium sulphate and aluminium sulphate in separate vessels. Mix the two solutions and shake the mixture. Examine the precipitate with a magnifying glass.

250. Alum. A double sulphate of potassium and aluminium has been formed, which has the formula

$$KAl(SO_4)_2 + 12\,H_2O.$$

This salt has been known for a long time as *alum*. In fact, the name of the metal aluminium was taken from this old name. It is almost the only salt of aluminium which crystallizes at all well, and for this reason it was for a long time almost the only compound of aluminium which was known in the pure state.

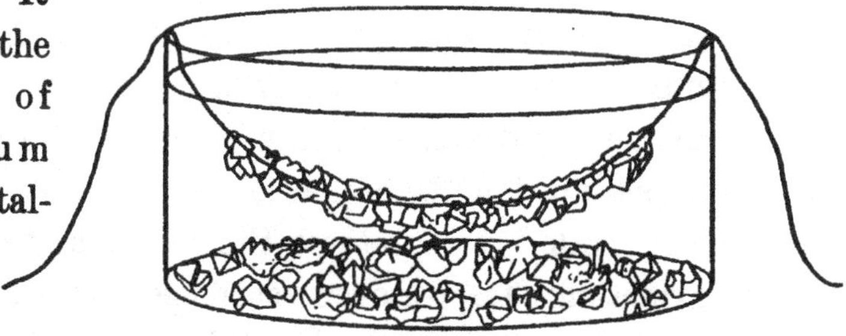

FIG. 85. Alum crystals

QUESTIONS

1. Explain why the development of electrical machinery has resulted in cheaper methods for the production of sodium, calcium, aluminium, and other metals.
2. In what two distinct ways is electrical energy used in chemical manufacturing?
3. Describe the process for making metallic aluminium.
4. Why is it that if aluminium combines so very readily with oxygen, things made of this metal do not immediately go to pieces?
5. What do you know of the basic properties of aluminium hydroxide?
6. Explain how aluminium hydroxide can act as an acid.
7. How do clays result from the weathering of rocks?
8. What is alum?
9. Compare the properties of aluminium hydroxide with the other hydroxides you have learned about.

PROBLEMS

1. What percentage of aluminium is contained in alum?
2. What percentage of water of crystallization does alum contain?

CHAPTER XXVIII

COPPER

251. Occurrence. Copper has been known and used by man for a very long time, because it occurs in the earth in metallic form and in compounds from which the metal can be very easily produced. It is nearly a *noble* metal, for it resists the action of air and water, while most of the metals we have so far studied about are very easily attacked. Even iron rusts in the air, and still more rapidly in water. Copper is oxidized in moist air, but the action is a slow one and soon stops because the oxide layer protects the metal underneath from further oxidation. Copper even resists sea water, which none of the other common metals do, and it is used in coppering the bottoms of ships. It is used for coins and medals because it is so permanent that it lasts practically unchanged for many years.

252. Properties. Copper is a red metal, but its color is usually tarnished by the layer of oxide which covers it. The copper deposited from solution in Experiment 78 showed the true color of the element. It has a high melting point, 1065° (a little higher than gold), and it is very tough, though not very hard. It is a very good conductor of electricity, and great quantities are used in making wires for electric circuits. Whenever an electric current passes through a wire a part of it is always transformed into heat, and a corresponding amount of electrical energy is lost. This loss of energy is different for different metals, and if we compare a number of wires, all of the same size and length, but of different materials, the one which changes the least electrical energy into heat is said to be the best conductor. If copper and iron are compared, we would find it necessary to have the iron wire seven times as large as the copper one, to have the same loss in each. We say that copper conducts seven times as well as iron.

253. Noble metals. Copper is the first metal which we have studied which is not dissolved by dilute acids. This means that copper cannot drive out the hydrogen from an acid, and, in fact, the reverse is true. Hydrogen gas can reduce copper ion to metallic copper. Under ordinary conditions the process goes on so slowly that it is barely measureable, but it can be very easily catalyzed.

Experiment 164. Arrange apparatus with a piece of platinum foil suspended in a tube so that half of it is in copper sulphate solution and the other half in hydrogen. The foil should be ignited in the burner immediately before the experiment is begun. Mark the height of the hydrogen in the tube. Allow the action to go on for a day or two, and then observe the height of the hydrogen. Take out the foil and examine it.

The reaction is

$$CuSO_4 + H_2 = H_2SO_4 + Cu,$$

or, in ionic form,

$$Cu^{++} + H_2 = 2\,H^+ + Cu.$$

The metals divide themselves into two classes with respect to their precipitation by hydrogen. The more noble metals, which we are to study later, are precipitated from their solutions by hydrogen, while all the metals we have examined up to copper set hydrogen free from acids. It is possible to form a series, arranged in the order in which one metal can precipitate the others from solution.

Experiment 165. (1) Place a piece of copper foil in a solution of silver nitrate; (2) place a piece of zinc in a solution of silver nitrate; (3) filter off the blue solution of (1) and add a piece of zinc.

The reactions are:

$$2\ AgNO_3 + Cu = 2\ Ag + Cu(NO_3)_2,$$
$$2\ AgNO_3 + Zn = 2\ Ag + Zn(NO_3)_2,$$
$$CuNO_3 + Zn = Cu + Zn(NO_3)_2;$$

or, in terms of the reacting ions,

$$2\ Ag^+ + Cu = Cu^{++} + 2\ Ag,$$
$$2\ Ag^+ + Zn = Zn^{++} + 2\ Ag,$$
$$Cu^{++} + Zn = Zn^{++} + Cu.$$

If we examine all the metals, we find that it is possible to arrange them in a series such that any metal precipitates those below it from solution, and is precipitated by all those above it. For the common metals we have this series: magnesium, zinc, aluminium, iron, copper, silver, gold.

254. The Daniell cell. We can now examine more closely the process in the Daniell cell, which served as a source of electrical energy in the experiments on electrolysis (Exps. 76, 77, and 78). The chemical reaction in this cell is a very simple one, as we have seen in section 106. Zinc goes into solution at the zinc plate and copper goes out of solution at the copper plate, just as when a strip of zinc is dipped into copper sulphate solution, except that in the cell we arrange for the zinc to go into solution and the copper to go out at two different points. In one case the chemical energy which is set free during the reaction appears in the form of heat, and this is where the two reactions, solution of the zinc and precipitation of the copper, go on at the same point. In the other case, where the two reactions go on at separate points, we are able to change the energy of the chemical reaction directly into electrical energy. In either case the new form of energy is produced at the expense of the chemical energy of the reaction. Another important point is that only those substances which can form ions when they dissolve in water can act in this way. It is evident that the properties of ions are determined by the presence of electrical energy.

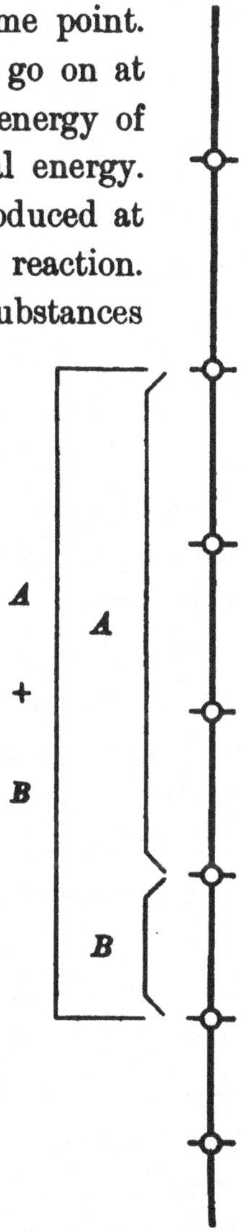

FIG. 86. Diagram of the series of metals

255. Electrical work. Any pair of metals, or in fact any reaction which can be so arranged that an oxidation and a reduction take place at different points connected by an electrolyte, can be made to produce electrical energy. Any pair of metals can be used in place of zinc and copper, but the quantity of electrical energy produced by the reaction of a given weight of any metal depends on the other metal which is used with it, and the reason for this lies in the way metals precipitate each other from solution. Zinc precipitates copper from its solutions; copper precipitates silver. If we make up two cells, one with zinc and copper as electrodes and the other with zinc and silver, we can get more work from the zinc-silver combination than from the zinc-copper one, the work in each case being calculated per gram of zinc.

We will get (per gram equivalent of each metal)

from zinc and copper the amount of work, A;

from copper and silver the amount of work, B;

from zinc and silver the amount of work, $A + B$.

There are two factors of electrical energy. One of these is quantity of electricity, and this is exactly proportional to the amount of material which dissolves at the anode and precipitates at the kathode during the time the cell is working (see Exp. 78 and sect. 108). The other factor is called difference of potential, and it is indicated by the position of the two metals in the series of section 253. If two metals are close together in the series, this factor is small, and so the quantity of energy produced when a gram molecule of each metal reacts is small, although the quantity of electricity which belongs to a gram molecule is constant. If the two metals are far apart in the series, their difference of potential is large, and a correspondingly large amount of energy is produced for the same weight of metal.

It is much like a water power. If you have the same weight of water, falling in one place from a height of ten feet and in another place from a height of twenty feet, twice as much work will be done by the same weight of water in the second case as in the first. The weight of the water is one factor in this, and the distance through which it falls is the other factor. Their product is gravitational energy, which can do work in turning a water wheel and running machinery.

When electrical energy is produced from chemical, if we choose any metal — zinc, for instance — as a standard, then the farther away from zinc a metal is in the series the greater the difference of potential (the greater the "fall") between zinc and the other metal, and the greater the quantity of

J. WILLARD GIBBS (1839-1903)

One of the broadest of general physical scientists. Applied general laws to the study of chemical equilibrium electrical energy which can be obtained from any given weight of zinc.

256. An electrochemical process. The purification of copper is carried on at the present time largely by the electrical deposition of the metal. A plate of pure copper forms the kathode, and impure copper in large billets forms the anode, both placed in solution of copper sulphate. The current is then passed through the cell so that copper goes into solution at the anode and deposits on the kathode in the form of pure metal, while the impurities remain in solution or fall to the bottom in solid form.

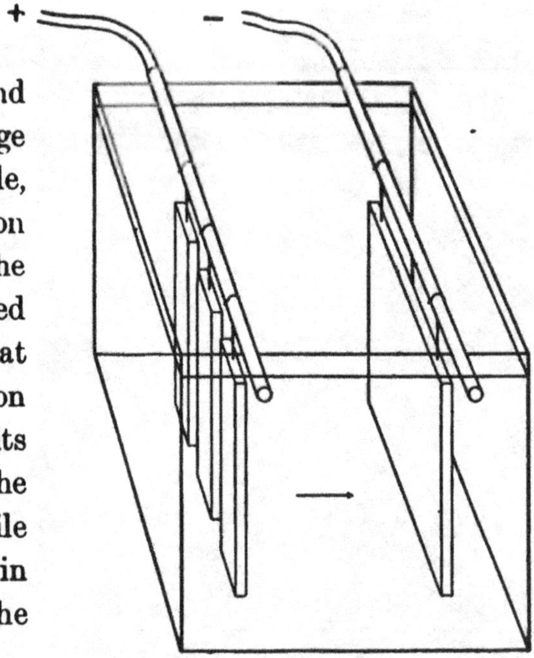

FIG. 87. Copper purification by electrolysis

257. Cupric hydroxide. This substance is formed by adding potassium or sodium hydroxide to a solution containing cupric ion, Cu^{++}.

Experiment 166. Precipitate cupric hydroxide as described. Heat the tube containing the precipitate and the solution from which it came.

The hydroxide breaks up very easily into cupric oxide and water;

$$Cu(OH)_2 = CuO + H_2O.$$

It is usually necessary to provide energy to produce water from a hydroxide, but in this case water forms with the production of a little energy, and so the process goes on easily of its own accord.

Cupric oxide and cupric hydroxide react with ammonium hydroxide solution.

Experiment 167. Add ammonia water to copper sulphate solution until the precipitate of hydroxide redissolves.

Cupric hydroxide is soluble in ammonium hydroxide solution, not because cupric hydroxide can act as an acid, but because a new ion of formula $Cu(NH_3)_4^{++}$ is produced, and copper ion is removed from the solution.

258. Cuprous ion. The cuprous ion is unstable in solution at ordinary temperatures and breaks up into cupric ion and metallic copper.

Experiment 168. To a solution of cupric chloride containing an excess of hydrochloric acid add some copper filings and boil. After the solution has cleared pour it into a large volume of cold water.

At the high temperature the reaction is

$$CuCl_2 + Cu = 2\,CuCl,$$

or

$$Cu^{++} + Cu = 2\,Cu^{+}.$$

Experiment 169. Boil some fine copper filings with strong copper sulphate solution. Filter hot and then allow the filtrate to cool.

259. Equilibrium. At high temperatures the reaction

$$Cu^{++} + Cu \rightleftharpoons 2\,Cu^{+}$$

goes toward the right and cuprous ion is formed. At lower temperatures it goes in the opposite direction and metallic copper is precipitated, with the formation of cupric ion. As long as metallic copper is present there will then exist a condition of equilibrium for each temperature, such that there is a definite proportion of cupric ion and a corresponding definite proportion of cuprous ion in the solution.

At ordinary temperatures the concentration of cuprous ion which belongs to this equilibrium is very small indeed, but at 100° C. it has a considerable value, as is evident from Experiment 169.

QUESTIONS

1. Why did the Bronze Age come before the Iron Age?
2. What are the properties of metallic copper?
3. Which is the stronger reducing agent, metallic copper or hydrogen? Explain.
4. Why does cupric hydroxide break up of its own accord into cupric oxide and water?
5. Why is cupric hydroxide soluble in ammonium hydroxide solution?
6. Tell what you can of the properties of cupric ion.
7. What can you state about the relations between cupric ion, cuprous ion, and metallic copper?
8. Where do you see a great deal of metallic copper every day?
9. Why is copper used for wires to carry an electric current rather than iron, which is much cheaper?

PROBLEM

1. What weight of silver is set free by 100 gm. of copper in Experiment 165? How does this compare with the relation between copper and silver found in Problem 1 at the end of Chapter XII?

CHAPTER XXIX

IRON

260. Value. Iron is the most important metal in the world at the present time, and the most valuable, not because its price per pound is high, but for the opposite reason. It is cheap and can be used for so many different purposes that it is practically a necessity to the human race.

Iron occurs native, that is, in the metallic state, only very rarely, but its ores, naturally occurring compounds from which it is made, are distributed all over the world. Nearly all rocks and fields are colored brown by the iron compounds in them, although the main constituents of soils are the silicates of aluminium, calcium, and magnesium, which are all colorless. Where plants have grown there are, of course, other coloring substances. These are the black and brown remains of the plants of former years.

The amount of iron in ordinary rocks and soils is too small to serve as a source of iron for technical purposes, but in many places there are great beds of iron ores, usually oxides of iron, and these contain such a large percentage of iron that it pays to make the metal from them.

261. The properties of iron. The metal itself is gray and tough, and *pure* iron is not used in any amount for technical purposes; cast iron, wrought iron, and steel, which are made up of iron and various amounts of carbon and other substances, being the commercially important forms.

Wrought iron is the nearest to pure iron, containing usually less than 1% of carbon, and it has the property of softening when heated, long before it begins to melt. On this property depends the process of *welding*. The smith heats to a white heat the two pieces of iron which are to be joined, and adds a substance which combines with oxide of iron to form a liquid *slag*. This removes the rust from the surfaces of the two pieces, and when they are brought into close contact by hammering, the clean surfaces of pure iron are united and the two pieces are *welded* together.

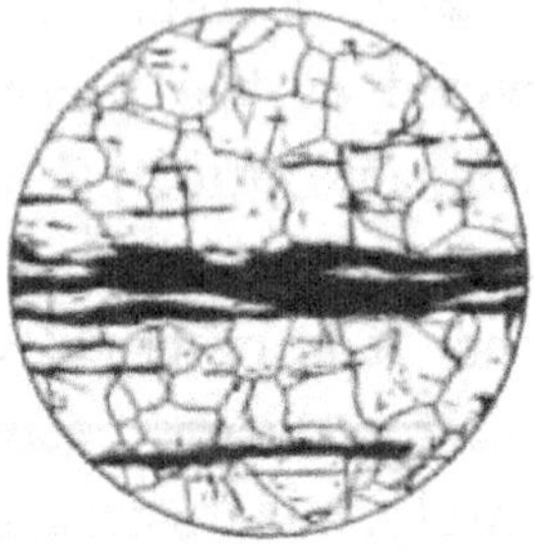

FIG. 88. Microphotograph of wrought iron

Cast iron is the cheapest form and it contains a much larger percentage of carbon than wrought iron, — sometimes as much as 6 or 7%. It cannot be welded, it is brittle instead of tough, and it melts much more easily than wrought iron.

FIG. 89. Microphotograph of cast iron

Experiment 170. Examine the broken surface of a piece of cast iron with a magnifying glass. Is it homogeneous, or can you distinguish more than one phase?

The dark spots in the cast iron consist of carbon in the form of graphite. In the process of making cast iron from iron ores it is heated with coal, and the molten metal which is formed dissolves carbon until the iron is saturated with it. Then when the metal cools it cannot hold all the carbon in solution, and it separates in the form of graphite when the mass solidifies. Part of the carbon is chemically combined with the iron, and when cast iron is treated with an acid, hydrocarbons are formed, which give an unpleasant odor to the hydrogen gas evolved.

Experiment 171. Place some pieces of cast iron in a test tube and add hydrochloric acid. What gas is produced? Try the odor.

When molten iron containing carbon is cooled very rapidly, the carbon has no time to separate as graphite, but remains combined with the iron, which becomes harder and more brittle even than ordinary cast iron. This form is called "white iron," or "spiegel," and it gives out a large amount of hydrocarbon gases when dissolved by an acid.

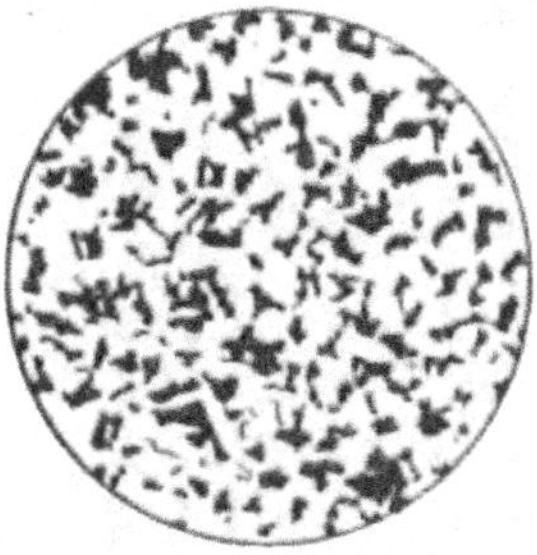

FIG. 90. Microphotograph of steel

Steel contains from $\frac{1}{2}$ to 2% of carbon, and it is more expensive than cast iron because purer material must be used in making it. Beside carbon, ordinary cast iron contains sulphur, phosphorus, and silicon, and these give iron its brittle properties. If perfectly pure oxide of iron and pure carbon could be obtained commercially, steel could be obtained directly, but all naturally occurring coals contain sulphur and phosphorus, and all iron ores contain silicon, so these elements are always present in commercial cast iron. *Wrought iron* is very nearly pure iron, from which the carbon has been burned out.

262. Iron ores. There are many ores of iron, the commonest of which are hematite (Fe_2O_3), limonite ($Fe_2O_3 \cdot Fe_2(OH)_6$), magnetite ($Fe_2O_3 \cdot FeO$), and siderite ($FeCO_3$). In order to obtain the metal from its ores the latter are first roasted in the air to drive off water and change them as far as possible into oxides, and also to make this raw material porous. The ores so prepared are heated with coal (or coke) and lime in a furnace.

263. The blast furnace. This is a great tower, often ninety feet in height and fifteen or twenty feet in diameter at the widest point, shaped like Figure 91. Air is blown in at the bottom of the furnace, and by the burning of the coal in the lower part a high temperature is produced and carbon dioxide is formed. This comes into contact with hot carbon farther up in the furnace and is reduced to carbon monoxide. This carbon monoxide takes the oxygen from the oxide of iron, reducing it to metallic iron, which combines with some carbon and is melted by the great heat as it passes down into the lower part of the furnace. Cast iron is drawn off at the bottom.

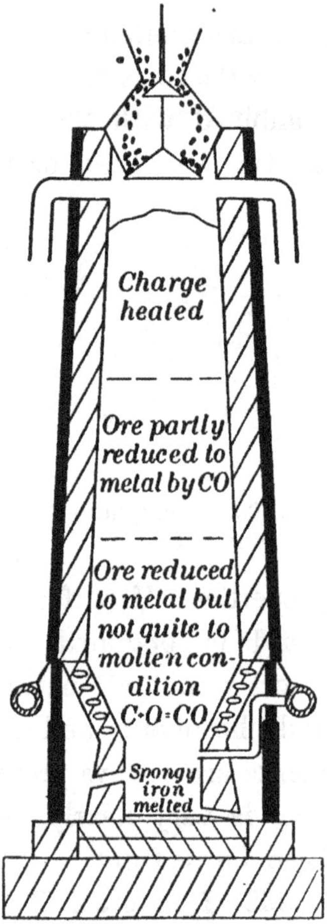

Fig. 91. The blast furnace

When the furnace has once been set in operation the process is a continuous one. Alternate layers of iron ore, coal or coke, and lime are dumped in at the top of the furnace from cars, and molten iron is tapped off at the bottom. The lime is necessary to form a *slag*, uniting with most of the silicon and other elements to form a molten mass, which collects above the iron near the bottom of the furnace, and which is tapped off from time to time through an opening above the one through which the iron is tapped.

264. Converters. In some plants the molten iron is run directly into large vessels called "converters," and changed into steel by burning out silicon, part of the sulphur and phosphorus, and part of the carbon.

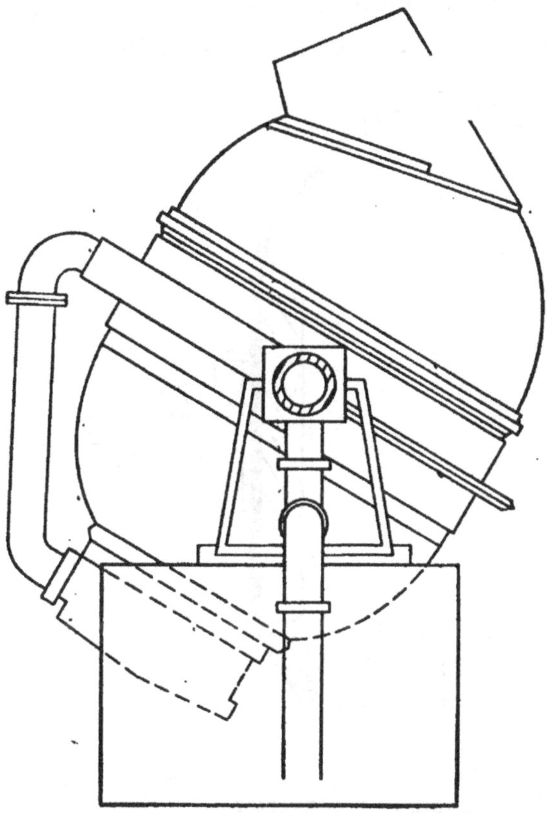

Fig. 92. Bessemer converter

In the Bessemer process the converter is a huge pot, shaped like Figure 92, supported on an axle so that it can be rotated, and provided with openings in the false bottom, through which a blast of air is blown (see section in Fig. 93). The converter is swung down into a horizontal position, and from one to twenty tons of molten pig iron is poured into it, either directly from the blast furnace or from a smaller furnace where pig iron is melted especially for the purpose. The air blast is then turned on and the converter is swung back to the vertical position. As the air is blown through the molten mass it combines with the carbon and silicon, and produces enough heat to keep the temperature above the melting point of iron. The carbon is oxidized to carbon monoxide, which burns at the mouth of the converter, and the silicon is oxidized to silicon dioxide, which is taken up in the slag. When the impurities have been burned out as far as possible, enough very pure spiegel is added to give carbon enough to produce

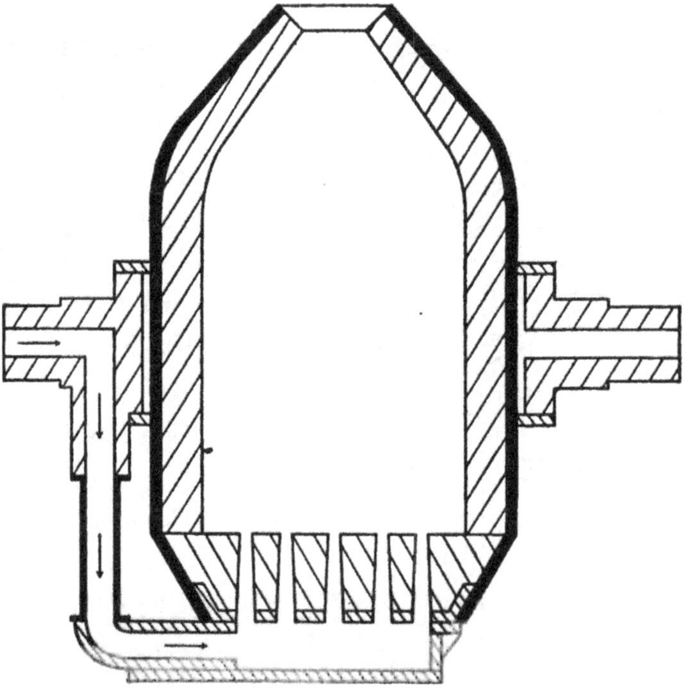

FIG. 93. Diagram of a converter

steel from the entire mass. The whole operation takes about twenty minutes, and then the contents of the converter is poured into molds, forming cast steel.

265. Basic lining. In the ordinary converter process sulphur and phosphorus are not completely removed from the steel, and it is left rather brittle. Iron of special purity, with a very minute quantity of these two impurities, must be used as the starting point for the process. There is, however, a modification of the original Bessemer process, which permits of the use of pig iron containing much larger amounts of sulphur and phosphorus, and which gives a tough steel of superior quality. For this process the converter is lined with a mixture of lime and magnesium oxide, all the other arrangements remaining the same as in the original process. The

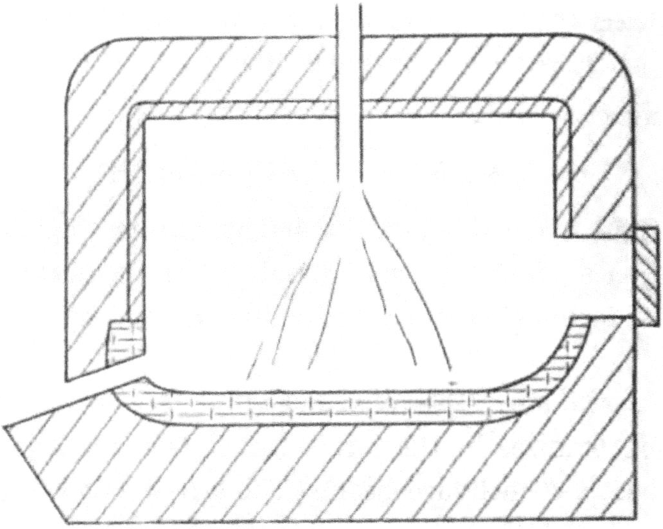

FIG. 94. Section through open-hearth furnace

phosphorus and sulphur combine with the basic substances of the converter lining to form phosphates and sulphates.

266. Open-hearth process. Other processes are used where steel having special properties is desired. In one of these, called the open-hearth process, a mixture of cast iron and wrought iron is melted in a shallow hearth with an oxidizing gas flame, that is, a flame containing an excess of air. This burns out the carbon slowly, and when the molten mass contains the right amount of carbon it is run from the furnace into a huge crucible and poured into molds. This process gives a tough steel of quality useful for construction purposes.

267. Ferrous ion. The solution left in the flask in Experiment 118, where we used ferrous sulphide and hydrochloric acid to make hydrogen sulphide, contained *ferrous* ion (Fe^{++}). This ion has a very slight greenish color and tastes like ink, which contains it.

Experiment 172. Add potassium hydroxide to solution of ferrous chloride. What are the properties of ferrous hydroxide?

The reaction is

$$FeCl_2 + 2\ KOH = 2\ KCl + Fe(OH)_2.$$

Pure ferrous hydroxide is white, but it oxidizes rapidly under the influence of the oxygen dissolved in the water of the solutions, forming another hydroxide of iron, which is dark in color.

268. Ferric ion. The substance which forms when ferrous hydroxide oxidizes in the air is called ferric hydroxide. If this is dissolved in hydrochloric acid, a new salt is formed,

$$Fe(OH)_3 + 3\ HCl = FeCl_3 + 3\ H_2O,$$

which is called *ferric* chloride.

The ion Fe^{++} can change to the ion Fe^{+++} by oxidation, and the substance which has the name ferrous chloride contains bivalent iron, while ferric chloride contains trivalent iron.

Experiment 173. Dissolve some iron filings or turnings in sulphuric acid. To a little of the solution add potassium hydroxide.

Add bromine water, a little at a time, to some of the solution of ferrous sulphate, keeping a slight excess of sulphuric acid present. After the solution has cleared, add potassium hydroxide solution.

Bromine is a good oxidizing agent, and in this case it has been used to oxidize ferrous ion to ferric ion, being itself reduced to bromine ion during the reaction.

The two ions of iron, Fe^{++} and Fe^{+++}, are very different in their properties. Ferrous ion is much like Mg^{++} in many of its properties, while Fe^{+++} is more like Al^{+++}.

Ferric hydroxide is a weak base, even weaker than aluminium hydroxide, so ferric chloride breaks up (hydrolyzes) into free acid and base.

Experiment 174. Dilute some ferric chloride solution until the yellow color is barely visible. Place some in two test tubes and heat one nearly to boiling while the other is kept cool for comparison.

Ferric hydroxide does not separate as a solid under these conditions, but remains in the colloidal form in the solution (see Exp. 145, on silicic acid). When the tube is cooled it does not immediately go into solution again, but after several days, solution takes place and the original color returns.

Colloids are peculiar substances which act in this way. The original condition of things is not restored in a colloidal solution when the temperature is brought back to its original point. Once the colloid has been precipitated it is often very difficult to get it back into the solution from which it came (see sect. 195).

Experiment 175. Test solution of ferric chloride with litmus. Is it acid or alkaline?

The conditions are very much the same as with aluminium chloride. The salt is hydrolyzed, and ferric hydroxide is so weak a base that it is removed from the solution, leaving the hydrogen ion to show its properties (see sect. 248).

Ferric chloride can be reduced to ferrous chloride by the addition of a substance which combines easily with chlorine.

Experiment 176. Add some magnesium powder to a solution of ferric chloride. After the reaction is over, filter, and test the filtrate for the ferrous ion by means of potassium hydroxide.

The reaction is

$$2\,FeCl_3 + Mg = 2\,FeCl_2 + MgCl_2.$$

We can use iron as a reducing agent just as well as magnesium. The reaction in this case is

$$2\,FeCl_3 + Fe = 3\,FeCl_2.$$

The oxides of iron which correspond to its two states of oxidation are ferrous oxide (FeO) and ferric oxide (Fe_2O_3). The first does not occur as an ore, but the second is hematite, the commonest of all iron ores.

269. Sulphides of iron. Iron forms two sulphides, one of which, FeS_2, is a very common mineral of yellow color and metallic luster, called *iron pyrites.* The other sulphide of iron, ferrous sulphide (FeS), can be made by heating pyrites, or by direct combination of iron and sulphur.

Experiment 177. Mix together fine iron filings and powdered sulphur in the proportion of their combining weights. Heat half of the mixture in a hard-glass test tube. Is heat absorbed or given out during the reaction?

Put the other half of the mixture in an evaporating dish with a little water and set it aside for a day or two. Then test both the substance in the dish and that in the test tube with hydrochloric acid. What has been formed in each case? (See Exp. 118.)

The same substance has been formed in both cases and the same amount of heat has been given out, calculated per weight of substance. But in one case the reaction went on rapidly, and accelerated itself by its own heat, while in the other case the reaction went on so slowly that the heat had time to escape from the mixture.

QUESTIONS

1. How does iron occur in nature?
2. In what forms is iron used for technical purposes?
3. How are two pieces of iron welded?
4. How is cast iron made?
5. Describe the operation of a blast furnace.
6. Explain the principle of the Bessemer converter.
7. Explain the principle of the basic lining of a converter.
8. What are the properties of ferrous ion? of ferric ion?
9. What do you know of the basic properties of ferric hydroxide?
10. Why are solutions of ferric salts acid to litmus?
11. How can you reduce ferric ion to ferrous ion?
12. What is a slag?
13. What is the composition of iron pyrites?
14. Write the equations for all the steps necessary to obtain each of the following compounds from metallic iron: ferrous chloride; ferrous sulphide; ferric chloride; ferric hydroxide.

PROBLEM

1. What weight of metallic magnesium is necessary to reduce 100 gm. of ferric chloride to ferrous chloride?

CHAPTER XXX

LEAD

270. Physical properties. Lead is another metal which has been known for a very long time because of the ease with which it can be produced from its compounds. It is a heavy metal, of density 11.4, with a rather low melting point, 325° C. The metal loses its luster very rapidly on exposure to air, covering itself with a thin layer of oxide, which is firm enough to prevent the metal underneath from further oxidation.

271. The action of pure water on lead. Whether or not a metal is attacked by water depends on the properties of the oxide layer which forms over it. Metallic lead is attacked to a much greater degree by very pure water than it is by ordinary water from the city pipes. The reason for this is that impure water contains carbonates and sulphates, and the carbonate and sulphate of lead are both difficultly soluble substances which form a firm, protecting layer over the metal. Pure water does not contain these substances, and the hydroxide of lead which it forms is not a firm, dense substance, and does not adhere to the lead and protect it. So while it is perfectly safe to use lead pipes for ordinary household affairs, it is not at all safe to use them for rain water or distilled water, because all the compounds of lead are very poisonous.

In the electrical series of the metals lead belongs between zinc and copper; that is to say, it precipitates copper from its solutions, but is precipitated by zinc. Lead has a place in the series very close to hydrogen, and this is one of the reasons why lead is not readily attacked by acids. Another reason is that most of the salts of lead are difficultly soluble, and another and still more important reason is the great difficulty with which hydrogen gas forms on its surface. Nitric acid attacks lead, but the action goes on only when the acid is fairly dilute. Lead nitrate is, of course, formed at first, but it is soluble in water and dilute nitric acid, and when this is used no protecting layer is formed and the action can go on unchecked.

Experiment 178. Make a saturated solution of lead nitrate by heating a solution containing an excess of the crystals and then allowing it to stand and cool down to room temperature. Add strong nitric acid to the clear solution.

272. The solubility law. There is a very general law here which states that every salt is less soluble in a solution containing one of its ions than it is in pure water. This is the reason why, in most analytical processes where we wish to produce a precipitate, we add an excess of the ion which is to form a part of the precipitated substance. In every case where the precipitate dissolves again in an excess of the precipitating agent, as was the case with aluminium hydroxide and copper hydroxide (see sects. 248 and 257), we may be sure that the dissolved substance contains an ion different from the one present in the original salt. In the case of aluminium hydroxide, which dissolves in an excess of potassium hydroxide, the ion AlO_2^{--} is formed, aluminium becoming part of the anion. In the case of the solution of cupric hydroxide in ammonium hydroxide solution, both copper and NH_3 are present in the new kation, $Cu(NH_3)_4^{++}$, which no longer has the properties of the copper ion.

273. Lead ion. The lead ion is colorless, as may be seen from its solutions. Lead sulphate is very slightly soluble in water, but it is much more soluble in hot water than in cold. Lead carbonate itself has no great importance, but the *basic* lead carbonate, which contains more lead than the ordinary carbonate, is a very important constituent of most paints, because of its *covering* properties. This word means that it takes only a small amount of a lead carbonate paint to produce a layer which does not permit any of the background to show through.

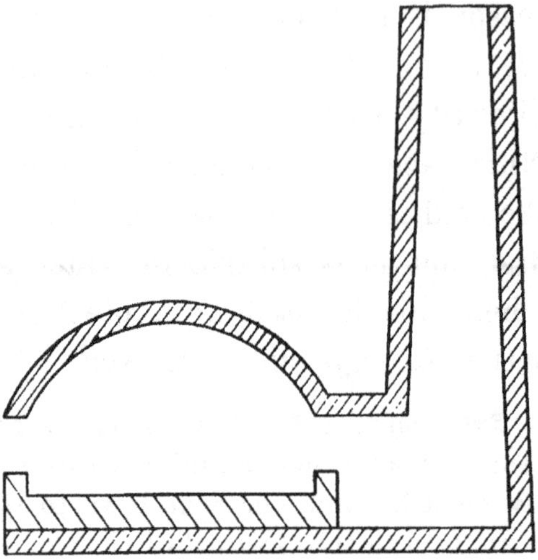

FIG. 95. The first step in the "air reduction" process for making lead

274. Lead sulphide. Lead sulphide is produced when hydrogen sulphide is passed into a solution containing lead ion. It occurs very widely distributed in nature as *galena*, and this ore is one of the most usual sources of lead. The process of reducing the sulphide to the metallic state is a very simple

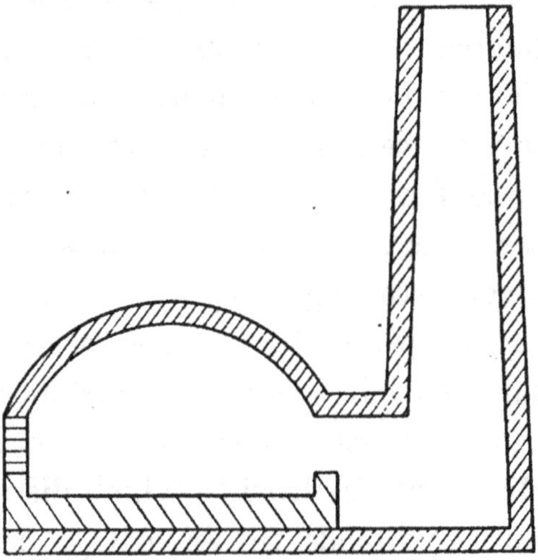

FIG. 96. The second step in the "air reduction" process

one. The ore is roasted until a part of its sulphur has been changed to sulphur dioxide and driven off as a gas, lead oxide being formed during the process. Instead of carrying the reaction to an end and producing lead oxide, afterward reducing this to metallic lead by means of coal, the mixture of lead sulphide and lead oxide resulting from the partial roasting is heated in a closed furnace. Here the reaction

$$PbS + 2\,PbO = 3\,Pb + SO_2$$

takes place, and if the preliminary roasting has been properly timed, complete reduction is the result.

This is a very good example of a technical method of reducing the cost of manufacture of a substance. Advantage is taken of the reducing power of sulphur; that is, its power of combining with oxygen.

275. Crude lead. Crude lead produced in this way contains a number of impurities, among the most valuable of which is a little silver, which is usually recovered by melting the pure lead and stirring metallic zinc into it. Zinc forms an alloy with silver, but not with lead, and the silver and zinc form a crust on the surface of the molten mass. This crust is skimmed off and the silver is recovered from it.

276. The oxides of lead. There are three oxides of lead, — PbO, *litharge*; Pb_3O_4, red lead, or *minium*; PbO_2, *lead peroxide*. All three are very useful substances, which find application in many places and for all sorts of purposes.

277. Paints. One of the basic carbonates of lead which has the formula $2\,PbCO_3 \cdot Pb(OH)_2$ is called *white lead*. It is prepared in very large amounts and is used as a pigment in white paints, and also as a body for paints which are to be colored with other substances. Most ordinary paints are merely suspensions of finely ground substances in a medium which is usually either oil or water. By the *covering power* of a paint is meant the power of a very thin layer to completely hide from view the surface beneath it. This covering power depends on rather complex optical properties, both of the suspended pigment and the medium which is used with it. White lead possesses covering power in a very marked degree, and this is the reason why it is so useful in the manufacture of paints. In order to be a useful pigment for ordinary purposes, a colored substance must possess certain definite properties. It must be cheap, it must be insoluble in the medium employed with it, and it must not decompose under the influence of light, air, and moisture.

QUESTIONS

1. What are the properties of metallic lead?
2. Why is lead more soluble in pure water than in water containing impurities?
3. Why can lead be used for ordinary water pipes, and why cannot lead pipes be used in stills from which pure water is to be produced?
4. What is meant by saying that a paint has good covering properties?
5. Describe a process for making metallic lead from lead sulphide.
6. How is crude lead freed from silver?
7. What oxides does lead form?
8. What other sulphates beside lead sulphate are difficultly soluble?
9. Could you use lead ion as a test for sulphate ion in a solution?
10. Explain Experiment 178.

PROBLEM

1. Calculate the percentage of lead in each of the three oxides of lead.

CHAPTER XXXI

MERCURY

278. Properties. This is the only metal which is a liquid at ordinary temperatures. Its melting point is – 39.4° C., and it boils at 358° C. under atmospheric pressure. It is a very heavy metal, of density 13.59, and it is a noble metal, since it does not oxidize in air at ordinary temperatures. Above 300° C. it does oxidize to form mercuric oxide (HgO), and this decomposes again at still higher temperatures into metallic mercury and oxygen.

Acids are not decomposed by mercury with formation of hydrogen gas. In fact, hydrogen can reduce mercury from the ionic to the metallic state. It has a place in the series of the metals below copper and not far from silver. It is attacked by nitric acid for the same reason that copper and silver are, and the reaction is of the same form as that for copper.

279. Ions. Mercury forms two ions, and therefore two series of salts corresponding to them. In one of these series mercury is univalent and in the other bivalent.

When mercury acts on nitric acid, and the metal is present in excess, mercur*ous* nitrate is formed. More nitric acid oxidizes the mercurous ion to mercuric ion, in the form of mercur*ic* nitrate. The two ions of mercury are very different in their reactions and in the solubilities of their salts.

Experiment 179. Add potassium hydroxide solution to solution of mercurous ion; to solution of mercuric ion.

280. Oxides. These two precipitates are not hydroxides, but oxides. Possibly the hydroxide may form at first in each case, but it breaks up so rapidly that neither mercurous nor mercuric hydroxide is known in the pure state (see sect. 256).

The mercuric oxide which is precipitated from solution of mercuric ion by hydroxide ion is yellow, instead of being red, like the oxide we used in preparing oxygen in Experiment 27. The difference is due simply to the different size of the grains in the two cases, and it is an example of a very general fact: substances which are red in large pieces change in color to orange and then to yellow as they are crushed finer and finer.

Experiment 180. Crush some pieces of potassium bichromate very finely by grinding in a mortar. Compare the color with that of some large crystals.

281. Calomel. The two chlorides of mercury have very different solubilities. Mercurous chloride is difficultly soluble, so when chloride ion and mercurous ion come together white mercurous chloride is precipitated. This is the substance which is commonly known as calomel, and its formula is $HgCl$.

282. Mercuric chloride. The most important of the mercuric salts is mercuric chloride, which is commonly called corrosive sublimate. This salt is not very soluble in water, but its solution is in the highest degree poisonous, and is used in medicine for sterilizing, since it destroys the dangerous organisms which might otherwise get into wounds or cause disease.

Mercuric iodide is formed when a solution containing iodide ion is added to one containing mercuric ion. If we use potassium iodide and mercuric nitrate, a result is obtained which suggests previous experiments.

Experiment 181. Add potassium iodide solution to mercuric nitrate solution, a little at a time, until the precipitate redissolves.

The precipitate which forms is HgI_2, and this dissolves in potassium iodide solution to form a salt of composition K_2HgI_4, which has ions K^+ and HgI_4^{--}. The mercury has here become a part of the anion, so it is no longer present in any amount as kation, and therefore no longer shows its original properties.

The mineral cinnabar is mercuric sulphide, HgS. It is a red substance, which is used in making fine red paints. Mercury and sulphur combine very easily, and mercuric sulphide can be made from the elements, even at ordinary temperatures, by rubbing them together in a mortar.

283. The metal mercury. Metallic mercury is made from cinnabar by heating the ore with iron in closed furnaces. The iron combines with the sulphur, and the volatile metallic mercury distills off and is condensed into its usual liquid form in cooled receivers. The metal is used in great quantities in the processes by which gold and silver are extracted from their ores. It is also a very important substance in chemical and physical laboratories. It is used for making barometers and thermometers, and for work with gases which are soluble in water. For this work it is used in a small pneumatic trough just as water is, except that trough and collecting vessels are smaller. Most of its uses in the laboratory depend on two facts: first, that at ordinary temperatures it is a noble metal, not oxidized by air; and second, that it is a liquid.

It will be seen from Figure 97 that the change of volume of mercury, which accompanies a change of temperature, is a very regular one. The curve showing the relation of these two properties is nearly a straight line. This is one of the reasons why mercury is so good a substance to use in thermometers. (Compare this curve with the corresponding one for water, Figs. 20 and 21.)

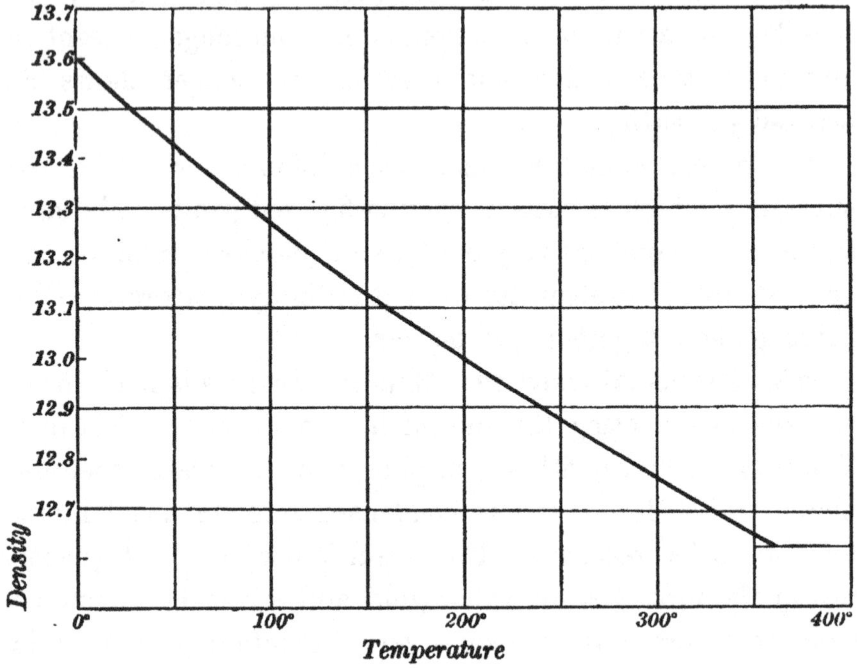

Fig. 97. Change of volume of mercury as the temperature is changed

QUESTIONS

1. What are the properties of the element mercury?
2. Write the reaction for the action of mercury on nitric acid.
3. What two series of salts does mercury form?
4. What are the properties of mercuric ion? of mercurous ion?
5. What is calomel?
6. What is corrosive sublimate?
7. What is the composition of the mineral cinnabar?
8. How is metallic mercury made from cinnabar?

9. Explain why mercuric iodide is soluble in excess of potassium iodide.

10. Why is precipitated mercuric oxide yellow instead of red?

11. Why is mercury so useful in the laboratory?

PROBLEMS

1. Calculate the percentage of chlorine in each of the two chlorides of mercury.

2. What percentage of metallic mercury is contained in pure cinnabar?

CHAPTER XXXII

SILVER

284. A precious metal. Silver is another very ancient metal, which occurs free in nature and which can be very easily reduced from many of its ores. It is a noble metal, not attacked by the oxygen of the air even at high temperatures, of white color and fine luster. It is a *precious* metal because of its qualities and because it does not occur in very large quantity. Calcium and silver are about equally expensive at the present time, but the reason for the price is an entirely different one in the two cases. Calcium occurs in enormous quantities in nature in the form of its compounds, but it requires a very large expenditure of energy to separate the metal in the free state. The cost of calcium is not in any way measured by its abundance, but only by the cost of manufacturing and the small demand for it. Silver, on the other hand, is very easily produced from its ores, but it occurs only in small quantities, and a large amount of material must be worked over before a pound of silver is produced.

Pure silver has great reflecting power when polished, and it is a good conductor of heat and electricity, — among the very best. These properties are made use of in the laboratory, but the pure substance has not many ordinary applications. It is too soft, and too easily scratched and bent. It is usually alloyed with about a tenth of its weight of copper, and the result is a hard alloy with all the good qualities of silver itself. The silver coins all over the world have this composition and silverware of the best grade has about this proportion of silver and copper.

285. Chemical properties. Silver does not act on acids with evolution of hydrogen gas. It lies far out at the other end of the series of metals beyond copper and mercury, and is therefore reduced by hydrogen even more easily than copper is. Nitric acid attacks it with formation of silver nitrate, the hydrogen which is formed reducing part of the nitric acid to nitric oxide and being itself removed from the reaction (see sect. 169).

Silver ion is colorless and very poisonous, and there is only one series of silver salts corresponding to the univalent ion Ag^+. Silver nitrate is the commonest of the silver salts, and is used to some extent in medicine. It forms large crystals, which are very soluble in water, and its solution is neutral to litmus paper, which shows that silver hydroxide is quite a strong base. It differs in this respect from the salts of most of the heavy metals, for they are usually hydrolyzed in solution (see sects. 248 and 268).

286. Silver oxide. The brown substance which is produced by the union of silver ion and hydroxide ion cannot be dried without losing water, and when dry it has the formula Ag_2O, silver oxide. This oxide decomposes on heating even more readily than mercuric oxide, oxygen escaping and leaving metallic silver behind.

Experiment 182. Prepare silver oxide by precipitation and drying in a drying oven at about 100° C. When it is quite dry, weigh out two or three grams carefully and determine the amount of oxygen it contains by collecting it over water and measuring its volume, using apparatus like that in Figure 98. Correct to standard conditions and calculate the weight of oxygen from the volume. Weigh the silver which is left. What is the atomic weight of silver?

287. Silver haloids. We are already acquainted with the haloid salts of silver. They are all difficultly soluble, and they are formed whenever Ag^+ and Cl^-, Br^-, or I^- meet in a

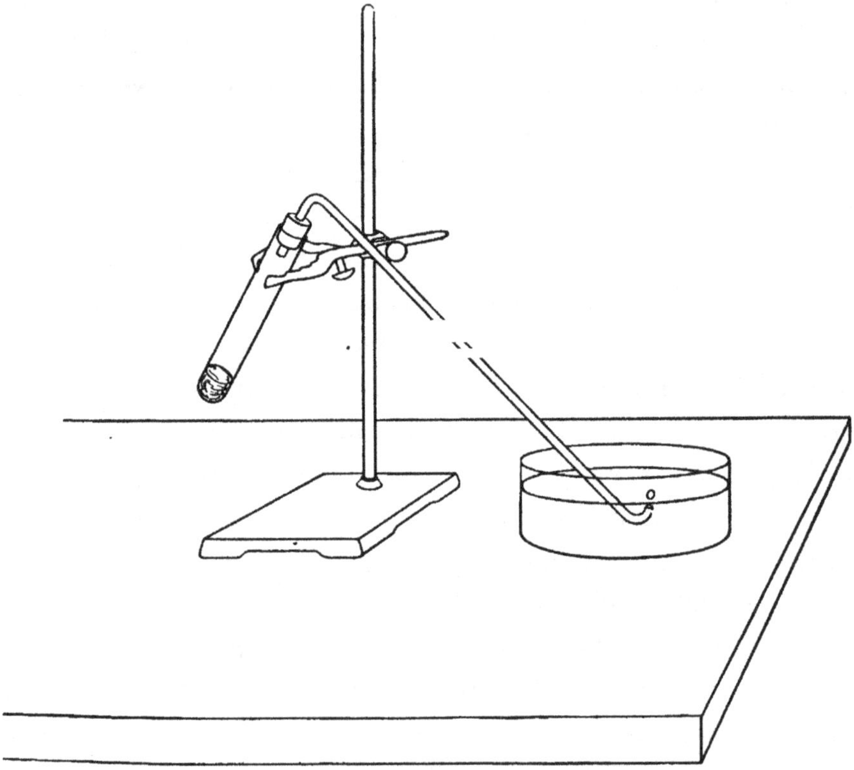

Fig. 98. Apparatus for determining the combining weight of silver

solution. They are stable salts and can be melted in the air without decomposition. They all occur in nature and are frequently produced in the laboratory as the result of tests for the silver ion or for the halogen ions.

Metallic silver can be easily recovered from the haloid precipitates by means of zinc. All that is necessary is to place a stick of zinc in the solution so that it touches the precipitate at the bottom of the vessel. In a short time all the silver will be reduced to the metallic state. The first action of the zinc is to reduce the silver which is in contact with it, and this forms a mass of metallic silver. The two metals then form a real galvanic cell, in which the chlorine ion goes from the silver chloride to the zinc, which is anode, while silver goes to the silver, which is kathode.

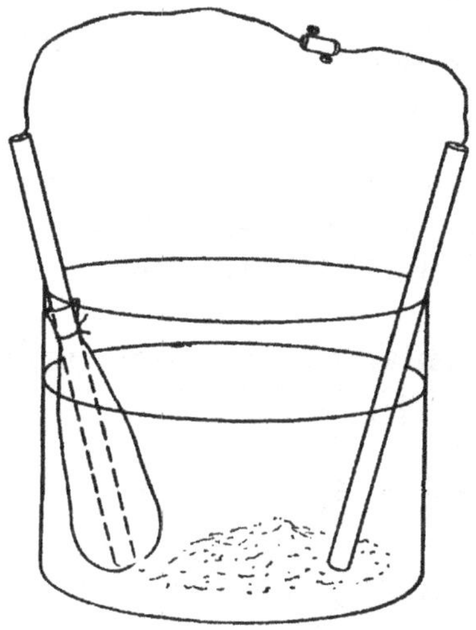

FIG. 99. Experiment to show the production of an electric current by oxidation and reduction

That an electric current is really produced under these circumstances can be shown in several ways. The zinc acts out in parts of the solution where there is no zinc, and this is a characteristic of an electric current.

Experiment 183. Wrap a piece of zinc in filter paper so that it cannot come in contact with the silver chloride in the glass. Dip this into the solution and connect it to a piece of silver or platinum wire which is placed in contact with the mass of silver chloride. To see whether a current is flowing through the wire, connect a sensitive galvanometer in the circuit.

288. Complex ions. The haloid salts of silver are soluble in some salt solutions. The chloride, which is what is called a very insoluble substance, is easily soluble in ammonium hydroxide solution. Under these conditions most of the silver ion must, of course, be removed from the solution. It has combined with the ammonia to form a new kation, $Ag(NH_3)_2^+$.

Silver chloride and the other silver haloids are soluble in sodium thiosulphate solution. In this case the silver ion is not removed from solution to become part of a new kation, but of a new anion (see sect. 292). Silver bromide is only very slightly soluble in ammonium hydroxide, but it is quite soluble in sodium thiosulphate. Silver iodide is not dissolved by ammonium hydroxide solution in measurable amount, and it is not as soluble in sodium hyposulphite as silver bromide is, but it is easily dissolved by potassium cyanide.

Experiment 184. Add potassium cyanide solution, a drop at a time, to a solution of silver nitrate, until the precipitate which forms has redissolved. Test the clear solution so formed for silver ion.

It is evident that the solution contains only a very small concentration of silver ion. The reaction is first

$$AgNO_3 + KCN = KNO_3 + AgCN,$$

and then

$$AgCN + KCN = KAg(CN)_2.$$

The potassium salt of a cyan-silver ion has been formed, and a solution of this substance contains even less silver ion than the solution of a silver haloid in sodium thiosulphate. It contains so little Ag^+ that silver iodide, which is very slightly soluble in ammonia water and not very soluble in sodium thiosulphate, is easily soluble in potassium cyanide.

289. Silver plating. A solution of potassium silver cyanide is used for silver plating because the silver which deposits from it is very smooth and not crystalline or spongy. The ions of this substance are K^+ and $Ag(CN)_2^-$, and when the current passes through the solution, potassium goes to the kathode and the silver goes to the anode as part of the anion $Ag(CN)_2^-$.

The precipitation of metallic silver at the kathode is a secondary process, which takes place according to the reaction

$$K + KAg(CN)_2 = Ag + 2\ KCN.$$

When articles are being plated the anode is made of a plate of silver, and there the reaction

$$Ag(CN)_2^- + Ag^+ = 2\ AgCN$$

takes place, with formation of silver cyanide, which dissolves in the excess of potassium cyanide of the solution, so that the bath does not change in concentration during the process of plating.

No K^+ can go out at the kathode, since it reacts with potassium silver cyanide as fast as it is reduced to potassium, and no CN^- goes out at the anode, since it reacts with the silver of the anode to form silver cyanide. So the final result of any amount of plating is the transport of silver from the anode to the kathode, just as in the simpler cases of electrolysis of copper sulphate and silver nitrate, which we studied in the chapter on Electrolysis.

290. Sensitiveness to light. The silver haloid salts have one very remarkable property, especially when they are in contact with organic substances. They are very sensitive to light and turn black after a short exposure to the sun. It is on this fact that the ordinary processes of photography depend.

The minute reactions of the photographic processes are difficult to follow, since they depend on changes in complex organic compounds, but the general principles can now be understood without any difficulty.

The plates (which it is much easier to buy than to make) contain silver bromide suspended in a gelatin film in the form of fine particles. Under these conditions silver bromide is very sensitive to light, and all the processes of making the plate, as well as those which must be carried out after its exposure, are done in a weak red light, which affects ordinary plates only very slightly.

The plate is placed in the camera, so that a sharp image of the object to be photographed falls on it. The lens is uncovered for a time depending on the brightness of the light and the sensitiveness of the plate.

During the exposure no visible image is produced on the plate, and although a chemical action has taken place it is so slight that it cannot be detected either by weighing or by examining any of the other properties of the film. That a change has taken place is shown by the property which the film has acquired. It can be *developed.* Wherever light fell on the plate, silver bromide was reduced,

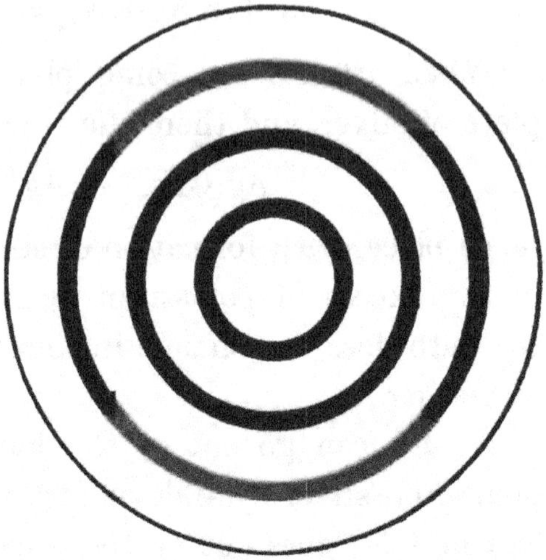

(*a*) Subject and positive

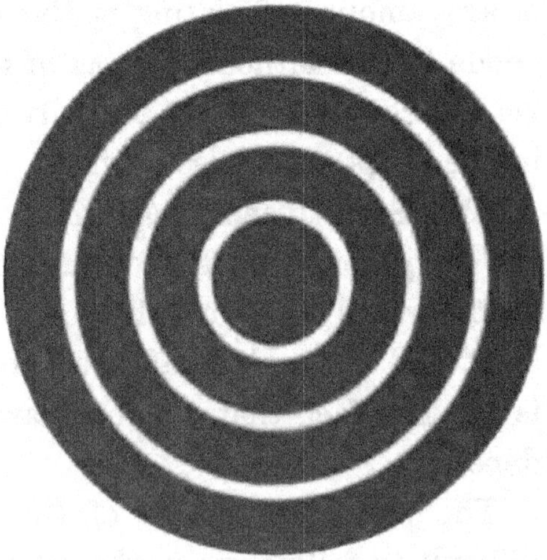

(*b*) Negative

FIG. 100

but whether to metallic silver or to a lower bromide (perhaps Ag_2Br) it is hard to say.

291. Development. However this may be, reduction has begun, and it can be carried on with the aid of *developers*, which are usually organic reducing agents. Under the action of the developer those parts of the plate which were exposed to light are reduced most rapidly, with the formation of metallic silver. The result is a *negative*, for where the light was strongest in the subject the plate is blackest, and where no light acted the plate remained white.

292. Fixing. In order that the plate may remain in this condition and not be further reduced by light, the sensitive silver bromide which has not been reduced must be completely dissolved out of the film, leaving only the black silver. A solution of sodium thiosulphate is used, which dissolves out all silver bromide and leaves simply clear glass in the parts of the plate where no light has acted, while the other parts of the plate contain silver in layers of varying density and thickness, depending on the intensity of the light which acted at each point.

The sodium thiosulphate which has been left in the plate by the fixing bath must be completely removed from the film by careful washing, otherwise it will gradually dissolve the silver itself and spoil the negative.

293. Positives. In order to produce a picture which looks like the original subject the same process must be done all over again. Copies are usually made on paper, and this is sensitized with a coating of silver bromide or silver chloride suspended in an organic medium. The negative is placed over the sensitive paper, so that all the light which reaches the film must pass through the silver layers of the negative.

Now the paper receives the most light at those points which were dark in the original subject, for the negative is nearly clear glass at those points. It receives the least light at points which were bright in the original, for the negative is covered with a thick layer of silver at these points. The result is a *positive,* and of course any number can be made from a single negative.

294. Printing processes. Modern printing papers are of two kinds, — developing papers and what are called printing-out papers. In the first the sensitive layer is like that of the plate; it contains grains of silver chloride or silver bromide suspended in gelatin, and the process of printing consists in exposing the paper under a negative for a short time and then developing the invisible image to the right degree of strength. In the other processes silver chloride is used, and another organic medium than gelatin may also be used. The paper is much less sensitive to light, and the action is allowed to continue until the print is at least as dark as is desired in the final result. It may then be simply "fixed" by dissolving out the excess of sensitive substances with sodium thiosulphate solution, or it may be "toned," to give it a more pleasing color than the one produced by merely fixing the print. The process of toning consists in replacing the silver of the print by gold or platinum, which give a better color to the picture. This is a simple process chemically, for gold and platinum are nobler metals than silver and further out in the series of the metals. So the silver goes into solution and the other metal is deposited in its place, replacing the color which the silver shows under these conditions by the color of the other metal.

QUESTIONS

1. Explain why it is that calcium and silver are about equally expensive.

2. Explain why silver is soluble in nitric acid.

3. What are the properties of silver ion?

4. What test have you used for silver ion?

5. Describe what happens when a piece of metallic zinc is placed in solution of silver ion.

6. Why is silver chloride soluble in ammonium hydroxide solution?

7. Why is it soluble in sodium thiosulphate solution?

8. Why are the silver haloids all soluble in potassium cyanide solution?

9. What does the order of solubility of the silver haloids in these three salt solutions tell you about their relative solubility in water?

10. What reactions take place during the process of silver plating?

11. Give a brief summary of the processes of photography, including exposure, development, fixing, printing, toning, etc.

PROBLEM

1. What is the percentage of silver in silver oxide? in silver iodide?

295. Properties. Although tin does not occur in the metallic state in the earth, it is easy to make from its compounds, and so became known very early in human history. It is a white metal with a low melting point, and in the series of the metals it lies between zinc and lead, nearer the latter than the former. It is therefore not a noble metal at all, but it is only very slightly acted on by air and water, because it covers itself with a very firm layer of oxide which stops further action. Enormous quantities of tin are used because of this very property. What is called tin plate is made by dipping iron vessels into molten tin, and this prevents rusting. Tin differs from lead in being just as resistant to pure water as to ordinary water, and it is therefore used in the laboratory for vessels in which the purest distilled water is condensed and stored.

296. Chemical properties. Since tin is so close to lead in the series of the metals it is not very rapidly attacked by dilute acids, and it only dissolves readily when the acid is heated. It forms two ions, one bivalent (Sn^{++}) and one tetravalent (Sn^{++++}), and two series of salts corresponding to them, which are called *stannous* and *stannic* salts, from the Latin name of the element, *stannum.*

Stannous ion is colorless, and its most usual salt is stannous chloride. Stannous ion is a very strong reducing agent, because the bivalent ion changes very easily into the tetravalent one.

Experiment 185. Add some mercuric chloride solution to a solution of stannous chloride and warm gently until no further change takes place. Collect the precipitate and rub it on a piece of clean copper foil.

Mercuric ion is reduced by stannous ion, first to mercurous ion (what is the white precipitate?) and then still further to metallic mercury. The gray color is an indication of the fine state of division of the metal. At the same time stannous ion is oxidized to stannic ion.

$$Sn^{++} + Hg^{++} = Hg + Sn^{++++}.$$

The crystals of stannous chloride have the composition

$$SnCl_2 \cdot 2\,H_2O,$$

and when this salt is dissolved in water the solution becomes turbid, in spite of the fact that stannous chloride dissolves rapidly. Sn^{++} is oxidized by the air to Sn^{++++} in small amount, and since the quadrivalent Sn^{++++} needs four anions (instead of two, as Sn^{++} does) to form a neutral salt, the solution becomes alkaline, and part of the Sn^{++++} is precipitated as hydroxide. In order to make a clear solution of stannous chloride it is necessary to keep a little metallic tin in the bottom of the bottle in which the solution is kept, and also to have the solution slightly acid with hydrochloric acid. (Why does this keep the solution clear?)

Sn^{++++} is not very well known. The hydroxide, $Sn(OH)_4$, acts much like an acid, and the chloride, $SnCl_4$, has none of the ordinary properties of a salt. It is a water-clear liquid, made by heating tin in a current of chlorine gas, and it fumes

J. J. BERZELIUS (1779-1848)

A leader in the chemical work of his time. First to use chemical formulæ in which the combining weight of each element is understood as expressed by its symbol strongly on exposure to air, reacting with the water vapor of the air to form hydrochloric acid and stannic hydroxide.

Cassiterite (SnO_2) is the only tin ore of much importance, and our whole supply of the metal comes from this ore.

QUESTIONS

1. What are the properties of metallic tin?
2. Why are iron vessels tinned?
3. What ions does tin form?
4. What can you say of the reducing power of stannous ion?
5. Why is it necessary to keep a solution of stannous chloride in contact with metallic tin and hydrochloric acid?
6. What are the properties of stannic chloride?
7. What is the principal ore of tin?

PROBLEM

1. What is the percentage of tin in cassiterite?

CHAPTER XXXIV

GOLD AND PLATINUM

297. Native metals. These precious metals are at the opposite end of the series of metals from sodium and potassium. It takes a great deal of work to prepare the latter metals from their compounds, and they take every opportunity of getting back into combination. Gold and platinum occur *native* (in the metallic state), and if they are combined with other elements they take every opportunity of getting back into the elementary form as soon as possible.

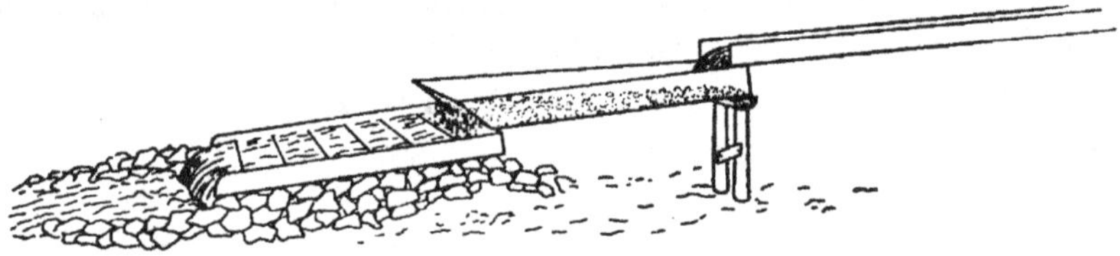

FIG. 101. A sluice for washing gold-bearing gravel

298. Gold. Gold is a rather soft, yellow metal of great density, malleable and ductile in high degree. While it finds occasional use in the laboratory, its principal uses are for making coins and jewelry. It seems to have advantages over all other metals as a basis for systems of coinage, because, in the first place, its production increases at a fairly stable rate, and second, because it has found no very important uses in the arts which could change its market value.

Platinum has all the physical and chemical properties which would make it even better than gold as a standard of value, but the great importance of platinum in manufacturing industries has so increased the demand that the price of platinum is now double what it was only a few years ago.

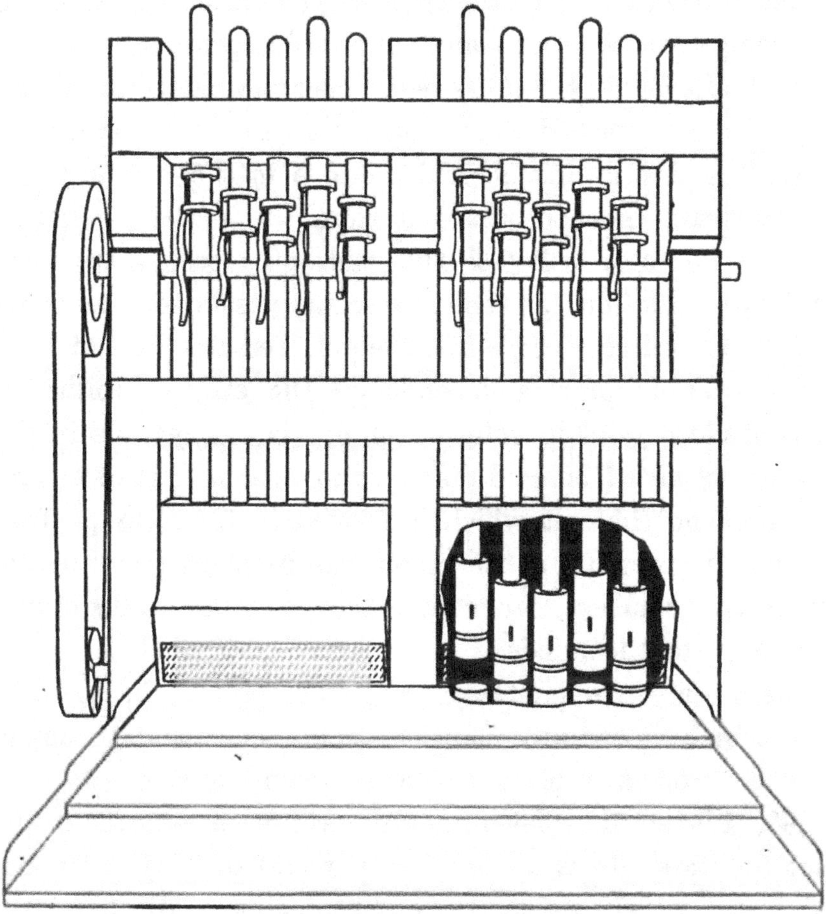

FIG. 102. Stamp mill for gold ores

The finely crushed material passes out with water through the screen and the gold is amalgamated with mercury on the plates

299. Gold milling. The older processes for separating gold from its ores depend on the high density of the metal (about 19). The finely crushed ore is washed until all the lighter constituents are carried away and the heavy grains of gold are left behind. A part of the gold is lost in this process, as the very finest grains are carried away with the dirt. At the present time other processes are used, the principal ones being the amalgamation process and the cyanide process.

Gold alloys very readily with mercury, and the finely crushed ore is washed over copper plates covered with mercury. The fine grains of gold dissolve in the mercury and the amalgam is then removed from the plates. When a quantity has been collected the gold and mercury are separated by distillation, the volatile mercury passing off and being condensed in a receiver, while the gold remains behind.

The cyanide process depends on the easy solubility of finely divided gold in solution of potassium cyanide in the presence of air. The gold dissolves to become part of a complex ion, and the salt which is formed is $KAu(CN)_4$. That the gold is present in the anion can be shown by electrolyzing the solution, when K^+ goes to the kathode and $Au(CN)_4^-$ to the anode.

The metal is precipitated from this solution by metallic zinc, which goes into solution, gold separating in finely divided form as a black or dark brown powder.

300. Platinum. Platinum also occurs in nature in the metallic state. It is a very heavy metal of gray color and fine luster, which resists the attack of almost all chemicals, so that it can be used in the laboratory for dishes and crucibles where porcelain and glass would be attacked. It never occurs pure in nature, but always alloyed with a number of metals which are very much like it, but which occur in still smaller amounts, and are therefore rarer metals than platinum itself. In order to purify it, it is dissolved in aqua regia, when the platinum goes into solution with the formation of an acid of formula

$$H_2PtCl_6,$$

platinum forming part of the anion $PtCl_6^{--}$.

Platinum is then separated from the impurities, some of which have gone into solution with it, by the addition of ammonium chloride, insoluble ammonium platinic chloride being formed.

Experiment 186. Wrap a piece of platinic chloride in a coil of thin platinum wire and heat it carefully, so that the platinum black which is formed adheres to the wire. Hold the mass thus produced in a strong stream of hydrogen gas coming from the hydrogen-generating apparatus.

See if you can light the gas at a gas jet by means of this finely divided platinum.

301. Catalytic power. Large quantities of platinum black, or platinum sponge, are used in the manufacture of sulphuric acid by the contact process, and in other catalytic manufacturing processes.

302. Aqua regia. Silver dissolves in nitric acid because the hydrogen, which would otherwise reduce it to metal again, is removed from the reaction (see sects. 169 and 285). Gold and platinum are insoluble in nitric acid because the very small amount of hydrogen which still remains uncombined is sufficient to reduce them to the state of metal. But if we arrange at the same time to remove the hydrogen by the oxidizing action of nitric acid, and to remove the metal by combining it to a complex ion as fast as it is formed, these noble metals can be dissolved. This is the function of aqua regia. The nitric acid attends to the removal of the hydrogen as far as possible, and the gold and platinum are taken into the complex anion and removed from the reaction in that way.

QUESTIONS

1. Explain why it is that gold and platinum occur native.

2. What are the properties of metallic gold? Why is it useful as a material for coins?

3. Why is platinum not so good as gold as a standard of value?

4. Tell what you can of the methods used to obtain gold from its ores?

5. What examples have you seen of the catalytic power of platinum?

6. Explain why gold and platinum are soluble in aqua regia.

CHAPTER XXXV

THE PERIODIC SYSTEM

303. The periodic system. So far we have used the combining weight as a property of an individual element, and only occasionally (sects. 136, 241) has any relation between the combining weights of different elements been suggested. It has been known for a very long time that similar elements show interesting relations connected with their combining weights, such as the ones which were found for the halogens when we summed up what we had learned about this group of substances.

It was noticed a number of years ago that many of the elements could be arranged in a series in such a way as to show many relations between the combining weights and the properties of the elements. These relations may be seen to the best advantage when all the elements are arranged in a single series in the order of their combining weights, beginning with the smallest and ending with the highest weight. On looking over such a series it is found that similar elements occur at regular intervals, and it is therefore possible to break up the long series into a number of smaller groups, each containing substances which show relations in their properties.

In the accompanying table this grouping has been made by placing related substances under each other. The long series is broken at fluorine (F, 19.0), and the next element, neon (Ne, 20.0), is moved back to form the first member of a

GROUPS

Periods	0 A	0 B	I A	I B	II A	II B	III A	III B	IV A	IV B	V A	V B	VI A	VI B	VII A	VII B	VIII
			H = 1.008														
1	He = 4.0		Li = 7.03		Be = 9.1		B = 11.0		C = 12.00		N = 14.04		O = 16.00		F = 19.0		
2		Ne = 20.0		Na = 23.05		Mg = 24.36		Al = 27.1		Si = 28.4		P = 31.0		S = 32.06		Cl = 35.45	
3	A = 39.9		K = 39.15		Ca = 40.1		Sc = 44		Ti = 48.1		V = 51.2		Cr = 52.1		Mn = 55.0		Fe = 55.9 Ni = 58.7 Co = 59.0
4				Cu = 63.6		Zn = 65.4		Ga = 70		Ge = 72.5		As = 75.0		Se = 79.2		Br = 79.96	
5	Kr = 81.8		Rb = 85.5		Sr = 87.6		Yt = 89		Zr = 90.6		Nb = 94		Mo = 96.0				Ru = 101.7 Rh = 103 Pd = 106.5
6				Ag = 107.88		Cd = 112.4		In = 115		Sn = 119.0		Sb = 120.2		Te = 127.6		I = 126.97	
7	Xe = 128		Cs = 132.9		Ba = 137.4		La = 139		Ce–Yb, etc., 140–173		Ta = 183		W = 184				Os = 191 Ir = 193 Pt = 194.8
8				Au = 197.2		Hg = 200.0		Tl = 204		Pb = 206.9		Bi = 208.5					
9					Ra = 225				Th = 232.5		U = 238.5						

new row. This brings sodium (Na, 23.05) under lithium (Li, 7.03). When all the known elements are collected in this way columns are produced, which are numbered 0, I, II, III, up to VII, and in each of these columns will be found the elements which are similar in their properties. It will be noticed that each column is subdivided into two columns, headed A and B, and that these are produced by writing the elements which occur in the column alternately to the right or the left. The elements in the half column, while showing in general the properties of the group, possess also special similarities in properties.

304. The groups. Group 0 contains elements which form no chemical compounds whatever. It includes argon and the other still rarer elements which are found in small quantities in the atmosphere.

Group I, Column A, contains the alkali metals. Column B is made up of those metals which are like silver in their chemical properties.

Group II contains the bivalent metals and includes both the alkaline-earth metals (see sect. 35) and the metals of the zinc group, which are in many respects similar.

Group III contains the earth metals, and Group IV is the group of the carbon elements.

So far in the series the groups contain almost wholly metals, and the basic properties are decreasing as we move towards groups of higher numbers. In the fifth group we find a good many nonmetals, although the elements of higher atomic weight, far down in the group, are more metallic in character. Finally, in Group VII we come to the halogens, which are wholly nonmetallic in character, and which have their place at the opposite end of the table from the alkali metals. The group numbered VIII contains metals which do not fit very well into the system. These are the elements which are similar to iron and platinum.

305. Properties expressed in groups. This table is an aid in remembering the properties of each element and the properties of its compounds. For example, we know that the formula for sodium chloride is NaCl. All the other elements of Group I form chlorides similar in their formulæ, — KCl, CuCl, RbCl, AgCl. In the same way the acids formed by the elements of the seventh group are each of formula similar to HCl.

If one member of a group has pronounced metallic properties, the others are usually metallic. If one member of a group has pronounced nonmetallic properties, the others will be like it in their behavior.

The physical properties of the members of a family usually show regular gradations as we go down the group. In the second group, for example, the densities of the metals in the second column are

Mg	1.75
Zn	7.00
Cd	8.67
Hg	13.6

The melting points are lower as we go down the group. They are

Mg	750° C.
Zn	420° C.
Cd	320° C.
Hg	– 39.5° C.

A similar gradation in the properties of the chlorine group has already been indicated by the table of section 136, and for the calcium metals in section 241.

306. Value of the system. It is quite evident that an arrangement of this kind brings out very sharply many of the similarities and differences in the properties of the elements. It is also evident that we have not yet discovered an exact law expressing relations between combining weight and properties. At some places in the table it is necessary to transpose two elements in order to make them fit the group. For example, in the cases argon-potassium and iodine-tellurium the element of higher combining weight must be placed before the other element in order to bring it into the group where it naturally belongs by reason of its chemical properties.

It is very striking that hydrogen has no place in the table. It should come at the top of the first column, as far as its combining weight is concerned, but its properties are very different from those of the elements of no combining power, for it is one of the most active of all the elements.

In spite of these exceptions the table is of great value. It will be found a great aid in remembering the properties of elements and their compounds, and chemistry is nothing more than just such knowledge. The table also enables us to predict to some extent, and if we know, for example, a good deal about the properties of the element sulphur, we know something about the other elements of the same group.

When this table was first made there were a number of vacant spaces in it which have since been filled. The whole first column, containing those elements which do not combine with anything, as far as we can find out, has been filled up. At least three metallic elements, — scandium, germanium, and gallium, — whose combining weights and properties had been predicted beforehand, have since been found, and they fit into their places in the table very well indeed.

CHAPTER XXXVI

ALLOYS

307. Alloys. Alloys are substances which are formed by melting together several metals. In some cases liquid metals do not dissolve one another (zinc and lead, for example), and when two such metals are melted together they separate again completely on solidification, the denser metal forming a layer at the bottom of the mass, with the lighter metal above it. In other cases the liquid metals dissolve one another to form a single liquid phase, and if separation occurs at all it occurs through the whole mass when it solidifies. Lead and tin act in this way. When metals are able to dissolve one another in the liquid state a remarkable mutual influence on their melting points may always be observed.

We have already learned that the melting point of a substance (and, of course, the freezing point as well) is lowered by dissolving another substance in it. This lowering of the freezing point is proportionate to the amount of the foreign substance dissolved.

Let us examine the facts in the case of lead and tin by means of a diagram (see Fig. 103). Along the horizontal line we will represent the constitution of the solution of lead and tin, and in these alloy diagrams it is found most convenient to express composition not in per cent by weight but in *equivalent per cent*, or, as it is also called, atom per cent. The equivalent weights of lead and tin are 206.9 and 119.0, and an alloy consisting of 206.9 grams of lead to 119.0 grams of tin would be described as containing 50 equivalent per cent (or 50 atom per cent) of each metal.

The constitution of our alloy is therefore to be measured in terms of this unit. At the left end of the line *AB* the point *A* indicates pure tin, while *B* at the other end of the

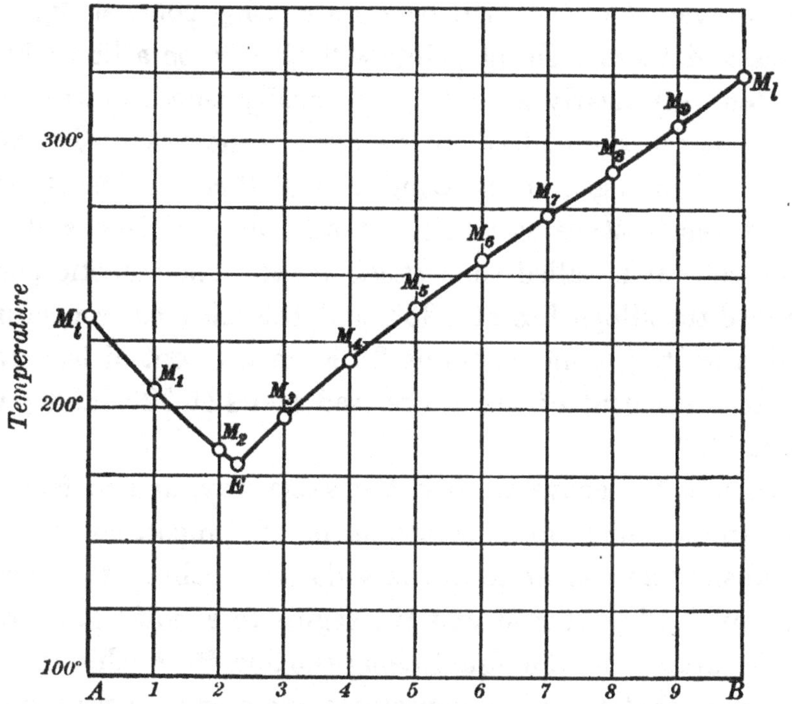

FIG. 103. Alloy diagram

line indicates pure lead. The point 5 at the middle of the line indicates that the solution contains *equivalent* parts of lead and tin. The vertical lines in the diagram represent by their length the temperature of the melting point of the solution. The point M_l represents the melting point of pure lead. The melting point of a mixture of nine parts of lead to one part of tin will be on the vertical line erected at the point 9, and it will be at the point M_9. As we add more and more tin we find a series of melting points (M_8, M_7, etc.), and these points will, in general, lie on a nearly straight line.

We can now do the same thing with tin, adding various proportions of lead to it. The melting point of pure tin will be represented by M_t, an alloy of nine equivalents of tin and one equivalent of lead will have its melting point at M_l, and so on, and these melting points will also lie on a line which is often very nearly straight and usually slopes downwards as the percentage of lead in the alloy is increased. The two lines will intersect at the point E, and this is evidently the lowest temperature at which a liquid alloy of lead and tin can exist. It is called the *eutectic point.* The eutectic point for lead-tin alloys lies at 183°, and the alloy corresponding to this melting point contains 23 atom per cent of lead and 77 atom per cent of tin. Pure lead melts at 325° and pure tin at 232°.

All metals behave in just the same way, and in fact all substances which form solutions in all proportions in the liquid state and separate in the solid state behave so. When the liquid alloy of lead and tin begins to solidify pure lead will separate from the liquid along the line M_l–E (this means, of course, that the alloy contains a large enough proportion of lead so that its melting point lies to the right of E). If the alloy contains sufficient tin so that its melting point lies to the left of E, then pure tin will separate from the alloy as it cools. At the point E the two metals separate in the solid state in the same proportions that they had in the liquid solution, and a mixture of solids corresponding to the point E will therefore have a constant melting point. There is another possibility in the case of alloys. The metals which are melted together may, and very frequently do, combine to form a new substance. This new substance will have a melting point of its own, which may lie lower or higher than the melting points of its components. The properties of this new compound will be different from the properties of either of its components. Such compounds of metals are usually metallic in their properties, but they may be very different in color, hardness, and so forth, from the metals used to produce them. Some alloys made from metals which conduct electricity very well are very much poorer conductors, and an alloy of several metals, none of which is measurably magnetic, may have strong magnetic properties.

Some alloys have very low melting points indeed, and technical application of this fact is very often made. Solders, for example, are alloys with rather low melting points. An alloy called Wood's metal is made by melting together 8 parts of lead, 45 parts of bismuth, 4 parts of tin, and 3 parts of cadmium, and this alloy melts at about 60°, although the lowest melting point of any of the metals used in it is 233°. By varying the constituents of an alloy like Wood's metal, substances having any melting point from 60° to that of the highest melting constituent can be obtained, and this fact is made use of in the construction of automatic fire alarms and safety plugs for boilers.

Brass is an alloy of copper and zinc. German silver consists of copper, zinc, and nickel. Type metal is an alloy of lead, tin, and antimony, and it owes its especial importance to the fact that it expands slightly at the instant of solidification, so that it can be cast very perfectly to fill all the fine portions of the mold. The alloys with mercury are called *amalgams*.

QUESTIONS

1. What would be the melting point of an alloy of 60 atom per cent of lead and 40 atom per cent of tin?

2. What two different alloys of lead and tin would have their melting point at 225°?

3. An alloy of lead and tin has its melting point at 290°. What is its composition?

4. An alloy consists of equal *weights* of lead and tin. What will its melting point be?

APPENDIX

TABLE I

DENSITIES OF METALS

Aluminium	2.6	Manganese	7.4
Antimony	6.6	Mercury	13.6
Arsenic	5.7	Molybdenum	8.6
Barium	3.7	Nickel	8.9
Bismuth	9.8	Platinum	21.5
Cadmium	8.6	Potassium	0.87
Cæsium	1.9	Rubidium	1.5
Calcium	1.6	Silver	10.5
Chromium	6.5	Sodium	0.98
Cobalt	8.6	Strontium	2.5
Copper	8.9	Tellurium	6.2
Gold	19.3	Thallium	11.8
Iron	(average) 7.8	Tin	7.3
Lead	11.4	Tungsten	19.1
Lithium	0.58	Vanadium	5.5
Magnesium	1.7	Zinc	7.1

TABLE II

DENSITIES OF NONMETALS

Boron	2.5	Sulphur	2.0 (allotropic forms)
Carbon	3. (allotropic forms)	Selenium	4.8
Iodine	5.0	Silicon	2.3
Phosphorus	1.3 (allotropic forms)		

TABLE III

DENSITIES OF GASES (in gms. ~~of~~ / lit)

(At 0° and 76 cm.)

Ammonia	0.762	Hydrochloric acid	1.628
Argon	1.782	Hydrofluoric acid	0.894
Bromine	7.143	Hydriodic acid	5.710
Carbon monoxide	1.251	Krypton	3.654
Carbon dioxide	1.965	Neon	0.893
Chlorine	3.167	Oxygen	1.430
Fluorine	1.697	Nitrogen	1.254
Helium	0.179	Water vapor	0.804
Hydrogen	0.090	Xenon	5.720
Hydrobromic acid	3.620		

TABLE IV

MELTING POINTS OF THE ELEMENTS

Aluminium	650°	Manganese	1250°
Antimony	625°	Mercury	– 40°
Barium	850°	Nickel	1500°
Bismuth	268°	Phosphorus	44°
Cadmium	320°	Platinum	1800°
Cæsium	26.5°	Rubidium	38°
Calcium	770°	Silver	955°
Cobalt	1600°	Sodium	97°
Copper	1065°	Sulphur (several allotropic forms with different melting points)	115°
Gold	1060°		
Iron (cast)	(about) 1200°	Strontium	—
Iodine	113°	Tellurium	450°
Lead	325°	Thallium	300°
Lithium	185°	Tin	232°
Magnesium	630°	Zinc	415°

TABLE V

BOILING POINTS OF ELEMENTS (At 76 cm. pressure)

Argon	− 186°	Hydrogen	− 253°	Sodium	800°
Bismuth	1450°	Krypton	− 152°	Sulphur	444°
Bromine	63°	Magnesium	1100°	Tellurium	1400°
Cadmium	775°	Mercury	357°	Thallium	1700°
Chlorine	− 33.6°	Nitrogen	− 194°	Tin	1500°
Copper	2100°	Oxygen	− 182°	Xenon	− 109°
Fluorine	− 187°	Phosphorus	290°	Zinc	925°
Helium	− 267°	Potassium	700°		

TABLE VI

BOILING POINTS OF COMPOUND SUBSTANCES

(At 76 cm. pressure)

Ammonia	− 33.5°	Sulphur dioxide	−8°
Hydrobromic acid	− 73°	Sulphur trioxide	46°
Hydrochloric acid	− 80°	Sulphuric acid (pure acid decomposes on heating)	
Hydrogen sulphide	− 62°	$+ H_2O$	340°
Mercuric chloride	300°	$+ 2H_2O$	190°
Nitric acid (the pure acid decomposes on heating)		Water	100°

TABLE VII

SOLUBILITIES

The solubility of a substance is definitely given only by a temperature-solubility curve similar to those of Figures 57 and 75, but the following statements may serve as a qualitative guide.

Bromides	Soluble except Ag, Hg, Pb
Carbonates	Not soluble except K, Na, NH_4
Chlorates	All soluble
Chlorides	Soluble except Ag, Hg, Pb
Hydroxides	Not soluble except alkalis and Ba
Iodides	Soluble except Ag, Hg, Pb
Oxides	Not soluble except Na, K, Ba (Ca)
Nitrates	All soluble
Phosphates	Not soluble except alkalis
Silicates	Not soluble except alkalis
Sulphates	Soluble except Ba, Ca, Sr, Pb

TABLE VIII

COMPOSITION OF EARTH'S CRUST

	Solid Crust Per cent	Ocean Per cent	Atmosphere Per cent	Total Per cent
Oxygen . . .	47.29	85.79	22.60	49.98
Silicon	27.21	—	—	25.30
Aluminium .	7.81	—	—	7.26
Iron	5.46	—	—	5.08
Calcium . . .	3.77	0.05	—	3.51
Magnesium .	2.68	0.14	—	2.50
Sodium. . . .	2.36	1.14	—	2.28
Potassium . .	2.40	0.04	—	2.23
Hydrogen . .	0.21	10.67	—	0.94
Nitrogen . . .	—	—	76.60	0.02
Chlorine . . .	0.01	2.07	—	0.15 (Chlorine and Bromine)
Bromine . . .	—	0.008	—	
Titanium. . .	0.33	—	—	0.30
Carbon	0.22	0.002	—	0.21
Argon, etc. . .	—	—	0.8	—
Phosphorus .	0.10	—	—	0.09
Manganese . .	0.08	—	—	0.07
Sulphur . . .	0.03	0.09	—	0.04
Barium. . . .	0.03	—	—	0.03
Chromium . .	0.01	—	—	0.01
	100.00	100.00	100.00	100.00

DIAGRAMS

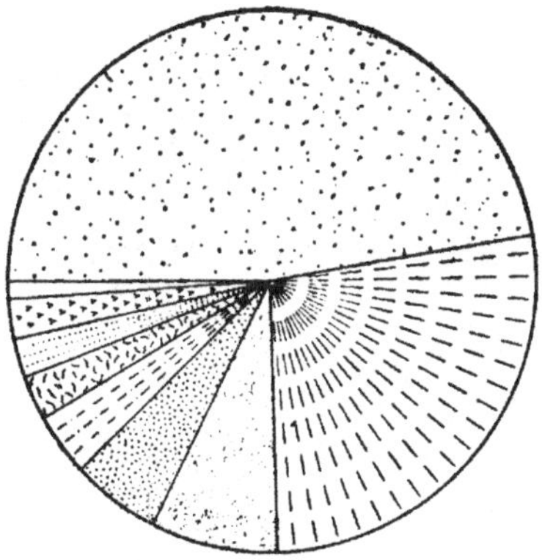

The solid crust of the earth

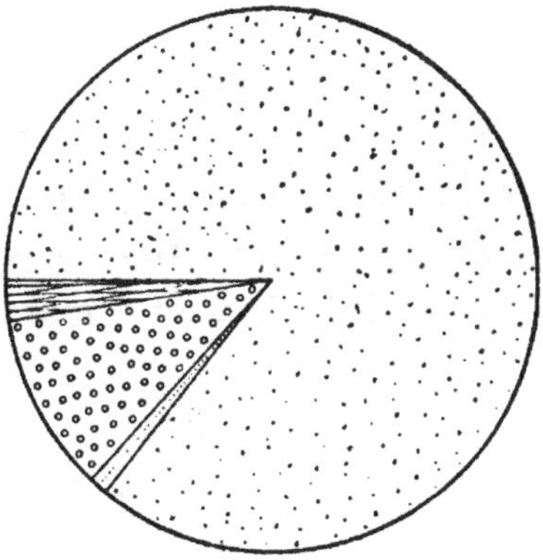

The ocean

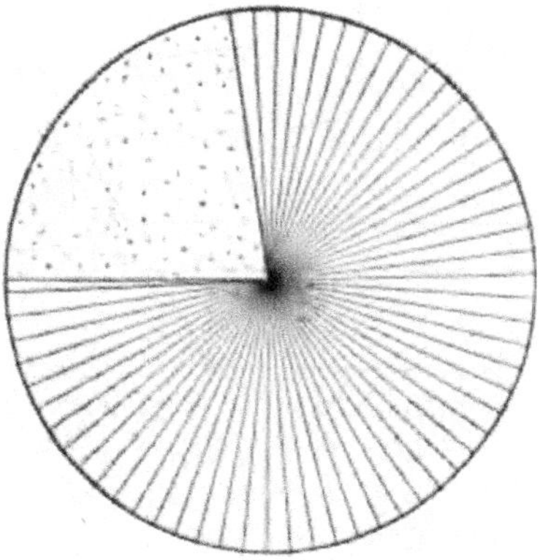

The atmosphere

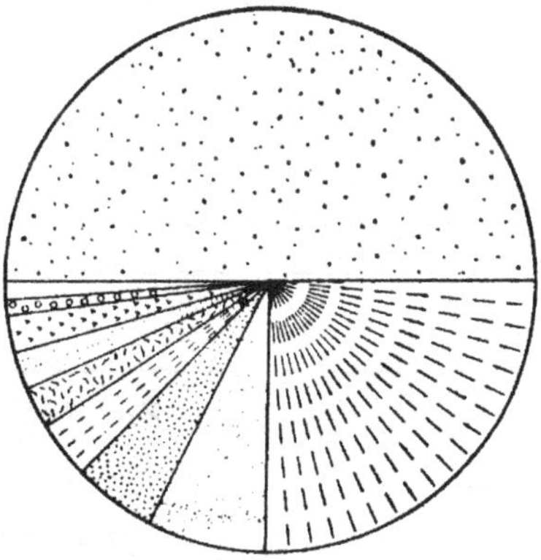

The whole crust of the earth

A GAS HOLDER

Several experiments (Exps. 50, 54, 82, 114, for example) require the aid of a *gas holder*. This is easily made from two large bottles by fitting them with stoppers and connections, as shown in Figure 104. A very little practice will make this a most useful piece of apparatus.

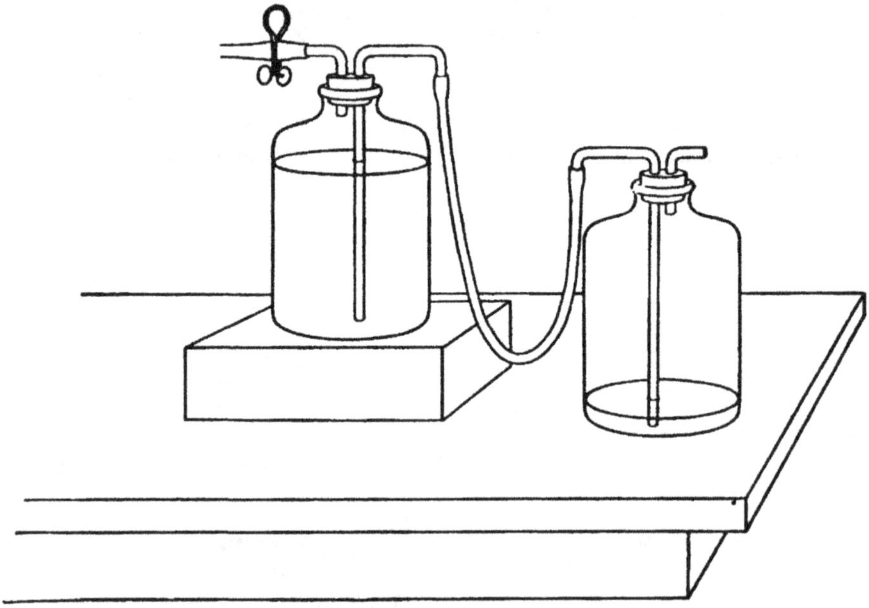

FIG. 104. A gas holder

STARCH PASTE

Starch paste is a delicate test for free iodine, and it is used in Experiments 103, 104, and 105. It is made as follows: Rub a gram or two of potato starch in a mortar with a little cold water until the grains are broken and a milky mixture is produced. Heat 200 ccm. of water to boiling and pour the mixture of starch and cold water into the boiling water, a little at a time, stirring constantly. This is the only way to make a clear, smooth paste from ordinary starch.

PLATINIZED ASBESTOS

In the modern contact process for making sulphuric acid (see Exp. 114), and in many other catalytic manufacturing processes as well, platinized asbestos is used. This is made by dipping clean asbestos into a rather concentrated solution of platinic chloride, or one of its salts, and then igniting over the Bunsen flame until the salt is completely reduced. Metallic platinum in a very fine state of division is formed throughout the asbestos.

INDEX

Made in the USA
Monee, IL
04 November 2026

67380787R10174